Urban Sustainability

The Urban Sustainability Book Series is a valuable resource for sustainability and urban-related education and research. It offers an inter-disciplinary platform covering all four areas of practice, policy, education, research, and their nexus. The publications in this series are related to critical areas of sustainability, urban studies, planning, and urban geography.

This book series aims to put together cutting-edge research findings linked to the overarching field of urban sustainability. The scope and nature of the topic are broad and interdisciplinary and bring together various associated disciplines from sustainable development, environmental sciences, urbanism, etc. With many advanced research findings in the field, there is a need to put together various discussions and contributions on specific sustainability fields, covering a good range of topics on sustainable development, sustainable urbanism, and urban sustainability. Despite the broad range of issues, we note the importance of practical and policy-oriented directions, extending the literature and directions and pathways towards achieving urban sustainability.

The series will appeal to urbanists, geographers, planners, engineers, architects, governmental authorities, policymakers, researchers of all levels, and to all of those interested in a wide-ranging overview of urban sustainability and its associated fields. The series includes monographs and edited volumes, covering a range of topics under the urban sustainability topic, which can also be used for teaching materials.

Rui Zhao

Multi-criteria Decision-Making Approaches to Sustainable Consumption and Production

Springer

Rui Zhao
School of Environmental Science
and Engineering
Southwest Jiaotong University
Chengdu, China

ISSN 2731-6483 ISSN 2731-6491 (electronic)
Urban Sustainability
ISBN 978-981-96-2774-5 ISBN 978-981-96-2775-2 (eBook)
https://doi.org/10.1007/978-981-96-2775-2

This Springer imprint is published by the registered company Springer Nature Singapore Pte Ltd.
The registered company address is: 152 Beach Road, #21-01/04 Gateway East, Singapore 189721, Singapore

Preface

In the context of "sustainable development", most 'materials' could be deemed to be 'hazardous' at some stage of their lifecycle, *i.e.* from extraction to final disposal. This book demonstrates multi-criteria decision-making (MCDM) applications can be used by engineers or policymakers to help limit environmental impacts, from the perspective of sustainable consumption and production, through the application of various computational approaches, by helping the decision-maker identify optimal solutions and mitigation strategies. In particular, the book aims to raise consumers' awareness of climate change and thus contribute to changing their lifestyles and purchasing behaviours. Such a change may increase the demand for more environmentally friendly products and promote sustainable production by manufacturer. The book further provides game theory analysis to model the likely behaviour of consumer, manufacturer and government in response to minimizing environmental impact, while improving performance along the three dimensions of the Triple Bottom Line (economic, social and environmental). It is essential reading for students, researchers, engineers, policymakers and those with a wider interest in sustainable development.

Chengdu, China
Rui Zhao

Acknowledgements This work was financially supported by the National Key Plan for Research and Development of China (2019YFC1905600), National Natural Science Foundation of China (41301639; 41571520), Sichuan Youth Science and Technology Innovation Team Funding (No. 2022JDTD0005), Science and Technology Research and Development Program of China Railway Group Limited (CRA 2022-Major-001; CRA 2023-Major-06), and Fundamental Research Funds for the Central Universities (No. 2682021ZTPY088).

Preface

Contents

Chapter 1
Sustainable Consumption and Production

1.1 Introduction

With the increasing depletion of natural resources and the threat of "climate change", the concept of sustainable development has been given more and more attention. Sustainable development is now broadly recognized as "the principles of the current process of economic and technological development to ensure that the use of environmental resources to satisfy present demands is managed in a way that they are not left so damaged or impoverished they cannot be used by future generations" [99]. The 'Triple Bottom Line' has been proposed to measure the impact of an organization's activities [49]. The Triple Bottom Line (TBL) has reflected the value of a company in terms of three dimensions, including not only its financial performance, i.e. profitability, but also its environmental and social resources [45]. Thus, a sustainable consumption and production should not only focus on the economic prosperity, but also pay much more attention to the environmental and social perspectives, ultimately benefiting all the stakeholders who will be affected by these three dimensions.

In this context, all materials can be considered as 'hazardous' in certain situations or at different stages of the product lifecycle [6], i.e. design, production, transport, use, and final disposal. In particular, the hazardous materials may be transformed into hazardous waste at the post-consumer stage. In addition to the immediate risk posed by 'hazardous' materials, it is also important in the socio-political context to attach importance to the 'carbon footprint', which may be an increasingly used indicator of environmental sustainability. Fossil fuels such as petrol, diesel, etc. have been widely consumed during production and transportation, resulting in the production of methane or other greenhouse gases. Since most waste, i.e. organic material, ends up in landfills, landfill gas, if uncontrolled, is considered to be one of the main sources of greenhouse gas emissions. Not only the reduction, reuse and recycling of waste materials, but also the provision of better, safer and more sustainable products for public consumption should be significant indicators in achieving sustainable consumption and production.

R. Zhao, *Multi-criteria Decision-Making Approaches to Sustainable Consumption and Production*, Urban Sustainability, https://doi.org/10.1007/978-981-96-2775-2_1

However, it is assumed that, depending on their business strategy, manufacturers will tend to focus primarily on their owned economic profit and would be reluctant to bear the additional economic costs of technological innovation to implement cleaner production. It is often assumed that manufacturers neglect their higher moral responsibility to provide 'environmentally sound' products with due regard to the Triple Bottom Line. For instance, product design typically still tends not to consider the full lifecycle, with little thought given to the post-consuming stage [19]. In order to provide 'better' and 'safer' products for the public, as well as to achieve long-term economic sustainability, it is the role of government to regulate the production process to ensure 'safe' products, not only in use to reduce their inherent risk, but also to mitigate their environmental impact. The question is whether government regulation is sufficient and whether manufacturers would be willing to bear the additional costs of green technological innovation.

Meanwhile, society should also focus on achieving sustainability of waste materials by transforming from "Cradle to Grave" to "Cradle to Cradle" to create a closed-loop system for zero waste, thus improving the product quality that is environmentally friendly [50]. "Cradle to Cradle" was firstly proposed by Walter R. Stahel in the 1970s, and subsequently developed by Dr. William McDonough and Dr. Michael Brauggart. Their joint publication "Cradle to Cradle: Remaking the Way We Make Things", demonstrates how the model can be achieved. This approach divides all materials into to two categories: 'technical' or 'biological' nutrients. Technical nutrients are those materials that are non-toxic, non-harmful, and have less environmental risk or negative impact on the natural environment. These types of materials can be reused directly in continuous cycles without loss of integrity or quality, or parts of their quality and functionality can be recycled, e.g. the cola bottle. Biological nutrients are organic materials that can be returned to the soil as fertiliser to provide food for microorganisms without harming the natural environment. Both nutrients are intended to prevent waste, reduce raw materials consumption, energy consumption, and environmental risk [65]. In this context, environmental considerations have played an important role in material selection, which is an increasing requirement in product design to provide better, safer and more sustainable products for public consumption.

In summary, as society faces major environmental challenges, it is possible to achieve sustainability by using fewer natural resources, optimizing recycling and reusing waste materials [67]. This book mainly focuses on the application of Multi-Criteria Decision Making (MCDM) for sustainable consumption and production, which is capable of providing mitigation strategies to limit environmental impacts and promote sustainable development. By using the system theory, game theory and optimization theory involved in MCDM, this book helps designers to identify environmental issues and provide better decision making for management and engineering practice to reduce environmental risks and the associated impacts, as well as to battle against climate change.

The book is divided into 6 chapters, in which Chapter 1 provides a brief background, discusses the problem situation, presents an overview related to the research topics, e.g. supply chain management, multi-criteria decision making, stakeholder's interaction, to provide insight into the study. Chapter 2 uses optimization theory, combined with novel underlying mathematical model, to identify the most appropriate strategy in the context of supply chain management, for manufacturers of commercial products to optimize the three dimensions of sustainability (TBL). Chapter 3 provides the fundamentals of game theory, to examine the strategic options selected by the involved stakeholders, including government and manufacturers, to promote cleaner production while maximizing the economic and environmental benefits. In order to promote sustainable production by reducing the consumption of 'hazardous' materials, Chapter 4 introduces approaches to materials selection for application by engineers to aid decision-making by taking environmental attributes into account. Chapter 5 uses the MCDM technique to minimize environmental risk regarding waste management. Chapter 6 draws conclusions from the previous chapters and discusses the research limitations, laying a foundation for further work. Future studies are proposed based on the concluding remarks and limitations.

1.1.1 Sustainable Supply Chain Management

Supply chain management (SCM) has been the subject of academic research since the early 1980s, covering a range of control and planning applications related to material selection, production, transportation, distribution, etc., as well as the potential collaboration between manufacturers, retailers and customers [11, 39, 43, 74]. With the rapid development of the global economy, the increasing demand for the consumption of commercial products has greatly accelerated the resource depletion and contributed to environmental pollution. Climate change has been identified as a critical factor that strongly influences the performance of supply chain management, and it may result in great uncertainty in terms of the public acceptance of various products [56, 85]. For instance, consumers' purchasing preferences may be influenced by the emissions information provided on a product label, especially as they become more aware of climate change [1, 111, 112]. This calls for the improvement of supply chain management towards a green transition (i.e. optimization based upon a lifecycle assessment) to strengthen its sustainability [31, 33].

Since the 1990s, increasing attention has been paid to optimizing supply chain management to take environmental issues into account has been paid increasing attention, including the issue of environmental investment recovery, internal redesign of the supply chain network, green coordination between upstream and downstream enterprises, and green initiatives [66, 83, 113]. Environmental investment recovery aims to encourage waste reduction and foster an attitude of reuse and recycle by means of reverse logistics and remanufacture [115]. Sheu et al. [84], proposed a linear multi-objective programming model to maximize the net profits of integrated logistics and the corresponding reverse logistics of used-products. The results indicated that the total net profits increased by 21.1%, compared to the existing operational performance. Furthermore, Yang et al. [103], developed a closed-loop supply chain network consisting of raw material suppliers, manufacturers, retailers, consumers and recycling centers. The variational inequality approach was used to obtain the optimal recyclability ratio of the used products and the conversion rate from recycled product to reusable material on the supply chain network. The same approach was used by Qiang et al. [77], who addressed the issue that consumers benefit from a lower market price when the expected yield rate of the recycled product increased. Kim et al. [54], proposed a model based on a stochastic model of the return lead-time of returnable transport items (RTIs) for deteriorating products, which identified that the return lead-time was positively correlated with the probability of delays in returning RTIs to the supplier. Moreover, excessive backorders would increase production and distribution costs.

The internal re-design of the supply chain network is mainly based on the implementation of cleaner production in the participating enterprises [90, 118]. Wang et al. [98], developed a multi-objective model to optimize environmental investment decisions using mixed-integer programming. With the objective of minimizing economic costs and carbon emissions, the best hierarchy of environmental protection was identified. A similar study was conducted by Abdallah et al. [1], who developed a mixed-integer programming model to minimize carbon emissions in the supply chain through green procurement. Mallidis et al. [62], built a strategic-tactical decision support model to optimize the design of ports of entry and transport modes. This model indicated that the shared use of transportation operations was effective in reducing the generation of carbon dioxide and particulate matter. Tseng and Hung [93] proposed a strategic decision model by considering both the operational and social costs of reducing carbon emissions. This approach showed that the social cost rate was negatively correlated with carbon emissions.

Green coordination between upstream and downstream enterprises includes green purchasing and green supplier selection [83, 118]. Zhao et al. [111], used game theory to select the optimal strategies for manufacturers in the supply chain to reduce their inherent risk and carbon emissions. Costs and benefits were identified as the main factors influencing the implementation of the selected strategy. Xie [102] proposed a model to examine how the level of energy conservation set by a local policy maker affected a manufacturer's energy saving and a retailer's selling price. Xie's work also argued that the decentralized chain should be given a threshold value for the energy saving level, while also considering the manufacturer's profit. Wu and Barnes [101] established a model that integrates ANP (Analytic Network Process) with MOP (Multi-objective Programming) to select appropriate partners to develop a green supply chain. The four clusters of the ANP network (cost, pollution control, quality, and resource consumption) were considered as the necessary criteria to evaluate the eco-efficiency of the partners.

With regard to the green initiatives of the consumers, Liu et al. [59] focused on the relationship between consumers' environmental awareness and the intensity of competition among green supply chain members. A two-stage Stackelberg game model was used to investigate the interactions between the different chain actors in order to identify different supply chain structures. In this effort, retailers and manufacturers involved in green operations were intended to benefit from the increased environmental awareness of consumers. Coskun et al. [16], proposed a goal programming model to re-design supply chain network based on consumers' green expectations. As the green consumer segment expanded, the retailer worked with suppliers to re-design the supply chain to meet consumers' expected level of green. A similar study was conducted by Ghosh and Shah [32], who further verified that green consumer markets provide better opportunities for supply chain stakeholders to launch green initiatives. Additionally, a cost-sharing contract might give rise to a higher greening of the supply chain network.

The creative management of a supply chain in the context of sustainable development, with the specific aim of minimizing the environmental impact that suppliers have on end users, is referred to as Green Supply Chain Management (GSCM) and has been discussed by many stakeholders [9, 80, 84, 88, 114]. Compared with conventional supply chain management, GSCM mainly focuses on the implementation of a green development strategy while managing both the external environmental pressures and the internal impetus of business innovation. GSCM seeks to develop options that can not only increase business profits, but also create environmental benefits, to ultimately achieve a 'win–win' situation [34, 55, 117]. Furthermore, the GSCM acts on waste reduction by encouraging the return of products to reduce raw material and energy consumption and prevent waste streams from entering the environment [36, 75, 111].

1.1.2 Stakeholders Involved in Sustainable Consumption and Production

Government and Manufacturers are the main actors in sustainable consumption and production. The Government role, is seen as promoting environmental protection to create a genuinely sustainable society, while ensuring economic development, which can be integrated into the DEFRA [18]. TBL provides an interpretation of the concept of 'sustainability' that, focuses not only on economic value, but also on environmental and social impacts that organisations have the power to influence [23].

Government aims to "encourage and drive innovation, research and design to improve sustainability throughout the lifecycle of all kinds of products" [44]. To be more specific, the main actions that government can take to intervene in the process of 'cleaner production' are to ban the sale of the least sustainable products in the market, to drive the existing market towards more sustainable products through regulation, and to encourage the development of 'better' and sustainable products through regulation, knowledge transfer, improved design and subsidies. As policies evolve, the whole product and market would become more sustainable over time. Furthermore, these actions mainly are enforced by using policy regulation, investing in environmental education for the public, funding for research and development for new technologies, research and the exploitation of knowledge, working with business on innovation in the environmental industries sector etc., to prevent or minimize environmental impacts and perceptional risks to the public, thus achieving cleaner production by providing "environmentally friendly products".

However, there are some dilemmas that affect government actions in the process of sustainable consumption and production. Primarily, the potential trade-off between economic performance and environmental impact should be considered at an early stage, as far as policy implementation difficulties are concerned. Other uncertainties include the dynamic behaviour of the system, e.g. the unintended consequences of regulation, environmental impacts, social responsibility, corporate behaviour, etc. It should be noticed that although there is a strong regulatory framework in place to drive industry towards sustainability, these inflexible regulations will only increase in number and scope if manufacturers do not respond [48]. Given the complexity, there is also the issue of "incentives or sanctions": how should government establish the reasonable incentive and restraint mechanism? For example, revenues from the landfill tax can be used to fund programmes to support business in improving resource efficiency [73].

For the manufacturers, there are three levels to running a business from bottom to top, which can be described as basic market demand (bottom line), a sustainable source of raw materials and energy for production, and the progressive achievement of 'green' societal value for environmental protection [48]. The manufacturer's role is to promote their individual firm's economic stability whilst acting in accordance with policy and norms of behaviour. In implementing sustainable production, manufacturers should recognize that they need to take more corporate social responsibility (CSR) for the public rather than just for their own profitability. For instance, some blue-chip organizations are using the Triple Bottom Line to manage their production processes. Moreover, corporate governance, risk management and control, business accounting and reporting can all be improved to varying degrees through the implementation of CSR [68].

However, manufacturers also face the dilemma that challenges and opportunities coexist with the risk and success, making it difficult to change their business strategies. Any change in strategy is an additional risk, but this can be offset by punitive measures, where manufacturers could be penalised for environmentally unfriendly production, for instance, higher carbon footprint of products, etc. The question is whether manufacturers should pre-empt regulatory action and seek to gain a competitive advantage or key selling point by considering the TBL. The traditional basis on which companies operate is governed by the profit motive, so there has been a tendency to neglect their higher moral responsibility to provide 'environmentally sound' and 'socially responsible' products with due regard to the Triple Bottom Line (TBL).

1.1.3 Dilemma in Sustainable Production and Consumption

Cleaner production, as a specific tool in the implementation of GSCM, aims to reduce the use of toxic materials and the amount of waste, ultimately mitigating the adverse impacts of the entire supply chain [60, 90, 116]. The following section examines the dilemma in management practice. From the standpoint of 'Hierarchy of Waste Management', cleaner production is proposed as an effective way to prevent environmental contamination and thus promote sustainable production. Cleaner production has been defined as "*the continuous application of an integrated, preventive strategy to processes, products and services to increase efficiency and reduce risks to humans and the environment*" [94]. It can be applied at any point in the life cycle of a product in order to mitigate the adverse impact on the environment and human health by reducing the amount and toxicity of raw materials used in the manufacture or packaging of a product; redesigning products in order to increase their durability, reusability and reparability; reforming the quality of products to minimize residual toxicity and waste when products reach the post-consumer stage; changing patterns of raw material demand and consumption that reduce the quantity and toxicity of waste generated, etc. [3,89].

While such strategies can bring both economic and environmental benefits, for example through energy savings or reductions in the number of raw materials purchased, or through reductions in waste disposal costs by reducing the amount and/or toxicity of waste to be disposed of [86], there is little dispute that the state has a role to play in providing a policy context that is conducive to cleaner production. For instance, the Chinese government has a stated policy of promoting cleaner production as a pathway to sustainable development, with subsidies, including tax breaks, equipment upgrades, etc., offered to enterprises for implementing cleaner production [15, 22]. As Smith et al. [86] also pointed out, a range of policies are needed to promote sustainable production.

However, policies will face uncertainties during the implementation of cleaner production, due to the conflict of interests between stakeholders, especially between government and manufacturers [22]. Furthermore, while environmental policies are essential for achieving sustainable development, they may not be sufficient. Even in the context of the EU, which has put environmental regulation firmly on the policy agenda since the 1970s, many manufacturers still primarily focus on the profit, which is the original bottom line for business [42], while not taking the full lifecycle of products into account and even not regarding waste management as a key business area. In addition, studies have shown that smaller organizations pay less attention to "waste prevention" or "cleaner production" than larger ones [24].

In order to assess the Cleaner Production Dilemma, this section develops a conceptual model using 'CATWOE' analysis to understand the main drivers, actors, etc. and influences on the ultimate environmental impact involved in a 'game'. The 'CATWOE' analysis is derived from the Soft Systems Methodology (SSM), which is assembled by the sequential process, Root Definition (RD) and Conceptual Model (CM). The root definition defines the purpose of the system, i.e. what the system is, while the conceptual model represents what the system should do, e.g. the activities which should be done in accordance with the purpose [8]. Moreover, the 'CATWOE' analysis selects the words from the RD and should be a test of the structure, providing a useful mnemonic that can be used to stimulate thinking about problems and possible solutions [14, 100].

There are six important elements of 'CATWOE', which can be specified as follows [14, 100]:

C—'Customers': the client, beneficiary or victim, i.e. who are the beneficiaries of the transformation process and how does the issue affect them
A—'Actors': the agents who would carry out or cause to be carried out the transformation process, i.e. who is involved in the situation, who will be involved in implementing solutions
T—'Transformation Process': the transformation of input into output, i.e. what processes or systems are affected by the issue
W—'World View': the perspective or taken-for-granted framework which makes the transformation process meaningful, i.e. what is the big picture and what are the wider impacts of the issue
O—'Owner': ownership of a system, control, concern or sponsorship, i.e. who owns the process or situation under investigation and what role they will play in the solution
E—'Environmental Constraints': environmental impositions, elements outside the system, interactions with the wider system, i.e. what are the constraints and limitations that will affect the solution and its success.

In this research, the root definition is deemed as "manufacturers, supervised by government, who provide environmentally friendly products to the public by reducing environmental risk and carbon footprint, and improving economic performance". The representative implication for 'CATWOE' is as follows:

C—Publics (members of society)
A—Manufacturers
T—Provide 'environmentally friendly products'
W—Reduce the environmental risk and carbon footprint and improve economic performance, followed by 'Triple Bottom Line' (TBL)
O—Manufacturers and Government
E—Not specified (including additional capital paid to upgrade existing equipment, technology, etc.)

This problem primarily arises from the product development in response to public demand and consumption. Moreover, it is important to note that commercial products are sometimes made with materials that are potentially 'hazardous', where the inherent risk can affect public health, and in some cases the health effects can be substantial. For instance, many materials have the potential to be toxic, especially carcinogenic, or to cause long term chronic health effects, such as formaldehyde which is a common substance used in some furniture. Accordingly, the public may be exposed to a limited risk from everyday consumer products. However, the public is prepared to accept this risk because of the utility of the products, with the balance between risk and benefit being a key element of subjective risk. In this context, it is important to note that a production decision-making process that does not take subjective risk into account may result in an inappropriate solution that will not be accepted by the public [19]. Therefore, the public demands 'better and safer' alternatives to consumer products. Meanwhile, manufacturers see huge commercial opportunities or advantages in improving their environmental performance. With 'green consumerism' gradually emerging into the mainstream market, government incentives and the consumer behaviour may become more rational. This would have created an unprecedented business opportunity for both "environmentally sound products" and services. In this case, public demand is deemed to be the 'principal driver' of cleaner production and supply chain management. In the Chapter 3, different stakeholders, i.e. Government and Manufacturers, with a large area of common interest are analysed to present their opportunities and dilemmas (Fig. 1.1).

1.1.4 Development of Multi-criteria Decision Making

With natural resources being exploited at an increasing rate, conflicts over resource extraction and allocation, pollution control and management are becoming more and more intense. Multi-criteria decision making (MCDM) has become an essential tool for analysing environmental issues, identifying appropriate indicators to better describe sustainable consumption and production, developing quantitative models to rank the possible alternatives in terms of their sustainable performance, and helping decision makers to make trade-offs among different alternatives [80, 81]. An overview regarding MCDM is provided to highlight its inherent advantage in provision of new insights and novel solutions with regard to sustainable consumption and production, which could identify appropriate strategies to optimize the three dimensions of sustainability in terms of supply chain management, adapt environmentally conscious design to promote cleaner production, and understand the drivers and actors involved.

According to the application scenarios, MADM can be divided into subjective and objective decision-making [58]. With regard to subjective decision making, it depends on the preference of the decision maker, which may lead to bias in the evaluation results [109]. The typical methods involved with their advantages and disadvantages are given in Table 1.1. AHP has been developed to assign weights to different criteria in a hierarchical indicator system [96]. Elimination Choice Translating Reality (ELECTRE) can further be coupled with AHP to rank alternatives according to their ranking relationships [95]. However, the information contained in the differences in ranking scores has not been explained [13]. The Preference Ranking Organization Method for Enrichment Evaluations (PROMETHEE) fills such gap by using different scores to indicate the degree of preference among alternatives [10]. AHP, ELECTREE and PROMETHEE are based on pairwise comparisons between attributes to rank alternatives, which is computationally expensive, especially when the number of alternatives or attributes increases in the decision making process [71, 78].

In order to minimize the uncertainty of result caused by the preferences of subjective decision-makers, objective decision making has been increasingly applied to

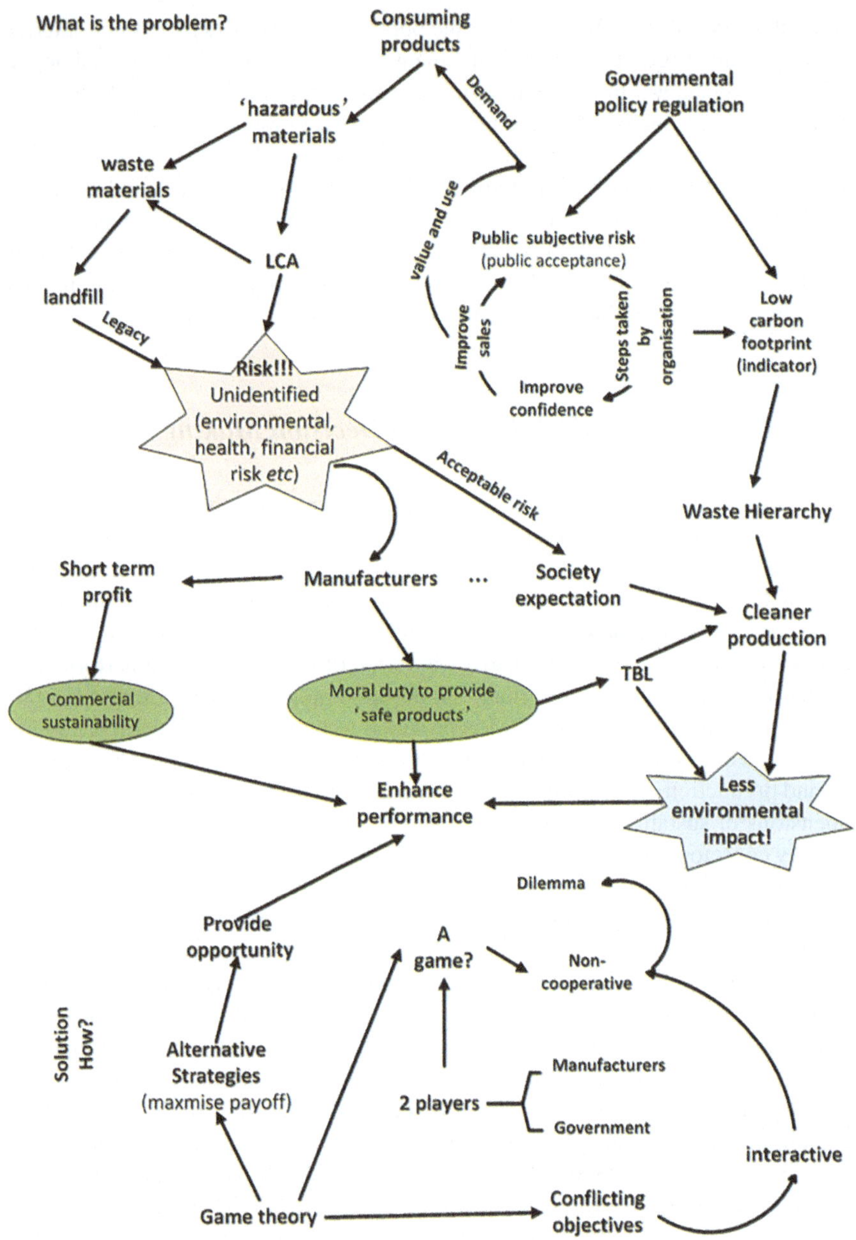

Fig. 1.1 Conceptual model of the cleaner production

Table 1.1 Comparison of typical subjective decision-making methods

Method	Advantage	Disadvantage
AHP	Easy to be understood by its hierarchy [63, 82, 91]	Computation becomes extremely complicated when a number of attributes are introduced [53, 69]
Fuzzy	Able to solve issues with attributes or indicators that are difficult to quantify [38, 46]	Subjective ranking tends to result in the polarization of indicator importance [106]
ELECTER	The pairwise comparisons are significantly decreased, since less input needed [21, 51]	Unable to deal with fuzzy information [2]

material selection. The typical methods involved, with their main advantages and disadvantages, are listed in Table 1.2. TOPSIS, VIKOR and COPRAS are comparatively simple, systematic and logically sounded for decision makers to obtain the optimal option when faced with a number of competing criteria [70]. The first two methods are similar but different from their normalizations in selected indicators [69]. VIKOR focuses on a compromise solution between the individual utility and the group utility, where the ranking of alternatives may differ due to the differences in subjective weightings [47, 108]. Similarly, COPRAS is identified by using the degree of utility to rank alternatives, which may lead to results that are sensitive to the normalization [7]. Technique for Order Preference by Similarity to Ideal Solution (TOPSIS) is more adaptable to qualitative and quantitative information, by discriminating alternatives in terms of their Euclidean distances from the ideal solution [27, 104]. It has been identified as reliable and efficient in a number of studies due to its ease of computation and clear trade-offs between different criteria [35]. In addition, TOPSIS is inherently effective when coupled with fuzzy to deal with uncertainty in material selection [26, 41]. Grey target decision-making has advantages in dealing with multi-attribute decision-making problems in the context of ambiguous decision environments [57]. It not only provides a ranking of the alternatives based on their advantages and disadvantages, but also enables the selection of the optimal option among feasible alternatives [4].

Besides, game theory has been widely used as a classic tool of MCDM to deal with multi-agent management problems, especially when there are conflicting objectives, to help decision makers improve the cooperative efficiency of the partners involved [12]. The solutions provided by game theory are usually arrived at by considering the interaction between the 'players' who are involved, which can be seen as a form of "interactive decision theory" [5]. It has been widely applied in various research fields, such as economics, marketing, production management etc. [87].

The specific process of game theory for decision making is described in Fig. 1.2, based on the above prototype. Within the loop cycle of the decision-making process, there are four key stages which are: (a) game situation analysis, (b) game modelling, (c) game result interpretation, and (d) decision-making for players in turn. The game situation is usually derived from a problem involving conflicts or different

Table 1.2 Comparison of typical objective decision-making methods

Method	Advantage	Disadvantage
DEA	The weightings for indicator are unnecessary, and computations can be implemented based on the raw data regarding the indicators [53, 110]	Limited input and output units, only able to determine whether the decision unit is valid [40, 92]
TOPSIS	No specific restrictions on data distribution, and information regarding raw data is retained as much as possible [104, 109]	Options may be close to both the positive and negative ideal solutions [105, 107]
VIKOR	Simultaneously considers the distance between each option and the ideal solution, and highlights the positional relationship of the optimal solution selected [47, 64]	Poor applicability for decision-making with scarce and fuzzy information [20]
Grey	Able to solve the decision issues with scarce information, and can provide valid results with limited available data [37, 97]	Lack of consistency test for indicator weightings [29]
PROMETHEE	Small computation load, able to achieve the partial or full ranking of alternatives [61, 76]	Indicator weightings cannot be given [17]
Preference Selection Index	Without the need to assign relative weights among attributes [25, 52]	Based on qualitative judgements [72]

interests [30]. In this premise, essential factors that formulate a game, such as "players", "payoffs" and "strategies" should be analyzed. The decision-maker with different interests can be seen as a 'player' in a game. A payoff is what a player may possibly gain in the game situation, depending on the strategic actions of all the players. Strategy can be set as the rule of a game, to describe how a player might play [28]. Moreover, questiones such as "how many players in the game", "how many strategies for each player", "what are the possible strategic actions for each player", "what are the possible payoffs for each player" etc., should be considered at this stage. Based on the analysis of the game situation, the practical problem can be quantified or semi-quantified in the appropriate models in terms of different game forms, e.g. normal or extensive form, or different game scenarios, e.g. static, dynamic, evolutionary game theory. The solution of a game is to determine what strategic action each player will choose, and the outcome can be interpreted as an equilibrium (agreement) by considering the interaction between the players to maximize the individual or group interests [79]. The agreement could be a long term or short term equilibrium to aid decision-making for the players involved. Once the equilibrium has been broken, a new game situation is generated to find out the alternative strategic action for each player.

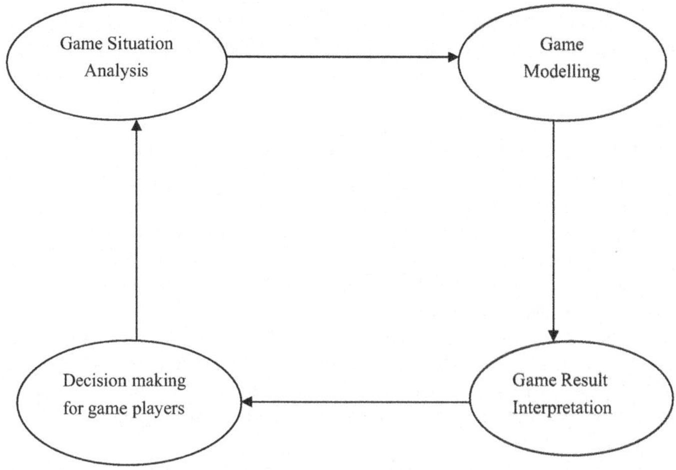

Fig. 1.2 Decision making process by using game theory

References

1. Abdallah T, Farhat A, Diabat A, Kennedy S (2012) Green supply chains with carbon trading and environmental sourcing: formulation and life cycle assessment. Appl Math Model 36:4271–4285
2. Akram M, Ilyas F, Garg H (2020) Multi-criteria group decision making based on ELECTRE I method in pythagorean fuzzy information. Soft Comput 24:3425–3453
3. Allen DT, Rosselot KS (1994) Pollution prevention for chemical processes: a handbook with solved problems from the refining and chemical processing industries. Hazardous Waste Research and Information Center, Illinois, pp 1–4
4. Ashby MF, Shercliff H, Cebon D (2018) Materials: engineering, science, processing and design. Butterworth-Heinemann, Oxford
5. Aumann RJ (2003) Presidential address. Games Econom Behav 45:2–14
6. Badurdeen F, Aydin R, Brown A (2018) A multiple lifecycle-based approach to sustainable product configuration design. J Clean Prod 200:756–769
7. Bairagi B, Dey B, Sarkar B, Sanyal S (2014) Selection of robot for automated foundry operations using fuzzy multi-criteria decision making approaches. Int J Manage Sci Eng Manage 9(3):221–232
8. Basden A, Wood-Harper AT (2006) A philosophical discussion of the root definition in soft systems thinking: an enrichment of CATWOE. Syst Res Behav Sci 23(1):61–87
9. Basu R, Wright JN (2008) Total supply chain management. Elsevier, Oxford, pp 245–257
10. Behzadian M, Kazemzadeh RB, Albadvi A, Aghdasi M (2020) PROMETHEE: a comprehensive literature review on methodologies and applications. Eur J Oper Res 200:198–215
11. Blanchard D (2007) Supply chain management: best practices. Wiley, Hoboken, pp 8–11
12. Cachon GP, Serguei N (2004) Game theory in supply chain analysis. In: Simchi-Levi SD et al (eds) Handbook of quantitative supply chain analysis: modeling in the e-business era. Kluwer Academic, London, pp 13–65
13. Chatterjee P, Athawale VM, Chakraborty S (2009) Selection of materials using compromise ranking and outranking methods. Mater Des 30:4043–4053
14. Checkland P, Scholes J (1999) Soft systems methodology in action. Wiley, Chichester
15. China NDRC (National Development and Reform Commission) (2008) Guidelines for the implementation of a cleaner production demonstration site program in China. http://www.chinacp.com/EN/PolicyDetail.aspx?id=40

16. Coskun S, Ozgur L, Polat O, Gungor A (2016) A model proposal for green supply chain network design based on consumer segmentation. J Clean Prod 110:149–157
17. de Almeida Filho AT, Clemente TR, Morais DC, de Almeida AT (2018) Preference modeling experiments with surrogate weighting procedures for the PROMETHEE method. Eur J Oper Res 264(2):453–461
18. DEFRA (Department for Environment Food and Rural Affairs) (2009) Guidance on how to measure andreport your greenhouse gas emissions. http://www.defra.gov.uk/environment/business/reporting/pdf/ghg-guidance.pdf
19. Deutz P, Neighbour G, McGuire M (2010) Integrating sustainable waste management into product design: sustainability as a functional requirement. Sustain Dev 18:229–239
20. Dev S, Aherwar A, Patnaik A (2020) Material selection for automotive piston component using entropy-VIKOR method. SILICON 12(1):155–169
21. Devi K, Yadav SP (2013) A multicriteria intuitionistic fuzzy group decision making for plant location selection with ELECTRE method. Int J Adv Manuf Technol 66:1219–1229
22. Dong XQ, Li CL, Li J, Wang J, Huang WT (2010) A game-theoretic analysis of implementation of cleaner production policies in the Chinese electroplating industry. Resour Conserv Recycl 54:1442–1448
23. Elkington J (1997) Cannibals with forks: the triple bottom line of 21st century business. Capstone, Oxford
24. European Environment Agency (2003) Europe's Environment: the third assessment. European Environment Agency, Copenhagen
25. Ezatpour HR, Torabi-Parizi M, Sajjadi SA, Ebrahimi GR, Chaichi A (2016) Microstructure, mechanical analysis and optimal selection of 7075 aluminum alloy based composite reinforced with alumina nanoparticles. Mater Chem Phys 178:119–127
26. Falqi II, Ahmed M, Mallick J (2019) Siliceous concrete materials management for sustainability using fuzzy-TOPSIS approach. Appl Sci Basel 9:3457
27. Feng YX, Zhang ZF, Tian GD, Fathollahi-Fard AM, Hao NN, Li ZW, Wang WJ, Tan JR (2019) A novel hybrid fuzzy grey TOPSIS method: supplier evaluation of a collaborative manufacturing enterprise. Appl Sci Basel 9:3770
28. Flesch J, Vermeulen D, Zseleva A (2017) Zero-sum games with charges. Games Econ Behav 102:666–686
29. Gagliardi F, Ciancio C, Ambrogio G (2018) Optimization of porthole die extrusion by Grey-Taguchi relational analysis. Int J Adv Manuf Technol 94(1–4):719–728
30. Geckil IK, Anderson PL (2010) Applied game theory and strategic behaviour. Chapman & Hall, London
31. Genovese A, Acquaye AA, Figueroa A, Koh SL (2017) Sustainable supply chain management and the transition towards a circular economy: evidence and some applications. Omega Int J Manage Sci 66:344–357
32. Ghosh D, Shah J (2015) Supply chain analysis under green sensitive consumer demand and cost sharing contract. Int J Prod Econ 164:319–329
33. Giudice F, Rosa GL, Risitano A (2005) Materials selection in the Life-Cycle Design process: a method to integrate mechanical and environmental performances in optimal choice. Mater Des 26:9–20
34. Gotschol A, Giovanni PD, Vinzi VE (2014) Is environmental management an economically sustainable business? J Environ Manage 144:73–82
35. Govindan K, Khodaverdi R, Jafarian A (2013) A fuzzy multi criteria approach for measuring sustainability performance of a supplier based on triple bottom line approach. J Clean Prod 47:345–354
36. Guide VDR, Souza GC, Wassenhove LNV, Blackburn JD (2006) Time value of commercial product returns. Manage Sci 52:1200–1214
37. Gupta M, Kumar S (2013) Multi-objective optimization of cutting parameters in turning using grey relational analysis. Int J Ind Eng Comput 4:547–558
38. Hafezalkotob A, Hafezalkotob A (2016) Fuzzy entropy-weighted MULTIMOORA method for materials selection. J Intell Fuzzy Syst 31:1211–1226

39. Harrison A, Hoek R-V (2008) Logistics management and strategy: competing through the supply chain, 3rd edn. Financial Times Prentice Hall, Harlow, pp 6–11
40. Hatami-Marbini A, Toloo M (2017) An extended multiple criteria data envelopment analysis model. Expert Syst Appl 73:201–219
41. He YH, Wang LB, He ZZ, Xie M (2016) A fuzzy TOPSIS and rough set based approach for mechanism analysis of product infant failure. Eng Appl Artif Intell 47:25–37
42. Henriques A (2004) CSR, sustainability and the triple bottom line. In: Henriques A, Richardson J (eds) The triple bottom line: does it all add up? Earthscan, London, pp 29–30
43. Hines T (2004) Supply chain strategies: customer-driven and customer-focused. Elsevier Butterworth-Heinemann, Amsterdam, pp 70–72
44. Hou GS, Wang Y, Xin BG (2019) A coordinated strategy for sustainable supply chain management with product sustainability, environmental effect and social reputation. J Clean Prod 228:1143–1156
45. Hussain N, Rigoni U, Orij RP (2018) Corporate governance and sustainability performance: analysis of triple bottom line performance. J Bus Ethics 149(2):411–432
46. Ishak NM, Malingam SD, Mansor MR (2016) Selection of natural fibre reinforced composites using fuzzy VIKOR for car front hood. Int J Mater Prod Technol 53(3–4):267–285
47. Jahan A, Edwards KL (2013) VIKOR method for material selection problems with interval numbers and target-based criteria. Mater Des 47:759–765
48. Kane G (2010) The three secrets of green business: Unlocking competitive advantage in a low carbon economy. Earthscan, London
49. Kannan D (2018) Role of multiple stakeholders and the critical success factor theory for the sustainable supplier selection process. Int J Prod Econ 195:391–418
50. Kausch MF, Klosterhaus S (2016) Response to 'Are Cradle to Cradle certified products environmentally preferable? Analysis from an LCA approach.' J Clean Prod 113:715–716
51. Kaya T, Kahraman C (2011) An integrated fuzzy AHP–ELECTRE methodology for environmental impact assessment. Expert Syst Appl 38(7):8553–8562
52. Khorshidi R, Hassani A (2013) Comparative analysis between TOPSIS and PSI methods of materials selection to achieve a desirable combination of strength and workability in Al/SiC composite. Mater Des 52:999–1010
53. Khoshand A, Rahimi K, Ehteshami M, Gharaei S (2019) Fuzzy AHP approach for prioritizing electronic waste management options: a case study of Tehran, Iran. Environ Sci Pollut Res 26(10):9649–9660
54. Kim T, Glock CH, Kwon Y (2014) A closed-loop supply chain for deteriorating products under stochastic container return times. Omega Int J Manage Sci 43:30–40
55. Kumar R, Chandrakar R (2012) Overview of green supply chain management: operation and environmental impact at different stages of the supply chain. Int J Eng Adv Technol 1:1–6
56. Lee KH (2011) Integrating carbon footprint into supply chain management: the case of Hyundai Motor Company (HMC) in the automobile industry. J Clean Prod 19:1216–1223
57. Li YR (2017) Study on scaling and corrosion of HDPE pipeline material in leachate transportation. Master of Thesis, Southwest Jiaotong University, Chengdu, China
58. Liu S, Chan FTS, Ran W (2016) Decision making for the selection of cloud vendor: an improved approach under group decision-making with integrated weights and objective/subjective attributes. Expert Syst Appl 55:37–47
59. Liu ZL, Anderson TD, Cruz JM (2012) Consumer environmental awareness and competition in two-stage supply chains. Eur J Oper Res 218:602–613
60. Madanhire I, Mugwindiri K, Mbohwa C (2015) Enhancing cleaner production application in fertilizer manufacturing: case study. Clean Technol Environ Policy 17:667–679
61. Maity SR, Chakraborty S (2015) Tool steel material selection using PROMETHEE II method. Int J Adv Manuf Technol 78:1537–1547
62. Mallidis I, Dekker R, Vlachos D (2012) The impact of greening on supply chain design and cost: a case for a developing region. J Transp Geogr 22:118–128
63. Mansor MR, Sapuan SM, Zainudin ES, Nuraini AA, Hambali A (2013) Hybrid natural and glass fibers reinforced polymer composites material selection using analytical hierarchy process for automotive brake lever design. Mater Des 51:484–492

64. Mardani A, Zavadskas EK, Govindan K, Senin AA, Jusoh A (2016) VIKOR technique: a systematic review of the state-of-the-art literature on methodologies and applications. Sustainability 8(1):37
65. McDonough W, Braungart M (2002) Cradle to cradle: remaking the way we make things. North Point Press, New York, NY
66. Mitra S, Datta PP (2013) Adoption of green supply chain management practices and their impact on performance: an exploratory study of Indian manufacturing firms. Int J Prod Res 52:2085–2107
67. Mohanty AK, Vivekanandhan S, Pin JM (2018) Composites from renewable and sustainable resources: challenges and innovations. Science 362(6414):536–542
68. Monagban P (2004) Put up or shut up. In: Henriques A, Richardson J (eds) The triple bottom line: does it all add up? Earthscan, London, pp 142–154
69. Mousavi-Nasab SH, Sotoudeh-Anvari A (2017) A comprehensive MCDM-based approach using TOPSIS, COPRAS and DEA as an auxiliary tool for material selection problems. Mater Des 121:237–253
70. Mousavi-Nasab SH, Sotoudeh-Anvari A (2018) A new multi-criteria decision making approach for sustainable material selection problem: a critical study on rank reversal problem. J Clean Prod 182:466–484
71. Mulliner E, Malys N, Maliene V (2016) Comparative analysis of MCDM methods for the assessment of sustainable housing affordability. Omega Int J Manage Sci 59:146–156
72. Noryani M, Sapuan SM, Mastura MT (2018) Multi-criteria decision-making tools for material selection of natural fibre composites: a review. J Mech Eng Sci 12(1):3330–3353
73. OECD (Organization for Economic Cooperation and Development) (2004) Addressing the economics of waste. OECD Publications, Paris
74. Oliver RK, Webber MD (1982) Supply chain management: logistics catches up with strategy. In: Christopher M (ed) Logistics: the strategic issues. Chapman & Hall, London, pp 63–75
75. Pan SY, Du MA, Huang IT, Liu IH, Chang EE, Chiang PC (2015) Strategies on implementation of waste-to-energy (WTE) supply chain for circular economy system: a review. J Clean Prod 108:409–421
76. Peng AH, Xiao XM (2013) Material selection using PROMETHEE combined with analytic network process under hybrid environment. Mater Des 47:643–652
77. Qiang Q, Ke K, Anderson T, Dong J (2013) The closed-loop supply chain network with competition, distribution channel investment, and uncertainties. Omega Int J Manage Sci 14:186–194
78. Rashidi M, Ghodrat M, Samali B, Kendall B, Zhang CW (2017) Remedial modelling of steel bridges through application of analytical hierarchy process (AHP). Appl Sci Basel 7:168
79. Romp G (1997) Game theory: introduction and applications. Oxford University Press
80. Rostamzadeh R, Govindan K, Esmaeili A, Sabaghi M (2015) Application of fuzzy VIKOR for evaluation of green supply chain management practices. Ecol Ind 49:188–203
81. Rostamzadeh R, Keshavarz Ghorabaee M, Govindan K, Esmaeili A, Nobar HBK (2018) Evaluation of sustainable supply chain risk management using an integrated fuzzy TOPSIS-CRITIC approach. J Clean Prod 175:651–669
82. Salwa HN, Sapuan SM, Mastura MT, Zuhri MYM (2019) Analytic hierarchy process (AHP)-based materials selection system for natural fiber as reinforcement in biopolymer composites for food packaging. BioResources 14(4):10014–10046
83. Sarkis J, Zhu Q, Lai KH (2011) An organizational theoretic review of green supply chain management literature. Int J Prod Econ 130:1–15
84. Sheu JB, Chou YH, Hu CC (2005) An integrated logistics operational model for green-supply chain management. Transport Res E-Log 41:287–313
85. Singh A, Mishra N, Ali SI, Shukla N, Shankar R (2015) Cloud computing technology: reducing carbon footprint in beef supply chain. Int J Prod Econ 164:462–471
86. Smith MH, Hargroves K, Desha C (2010) Cents and sustainability: securing our common future by decoupling economic growth from environmental pressures. Earthscan, London

87. Sun HJ, Wu JJ, Wang W, Zhang WY (2013) Production game of multifirms with different decision-making patterns. Proc Inst Mech Eng Part B J Eng Manuf 227(9):1393–1402
88. Tan Y, Shen L, Yao H (2011) Sustainable construction practice and contractors' competitiveness: a preliminary study. Habitat Int 35:225–230
89. Tchobanoglous G, Kreith F (2002) Handbook of solid waste management, 2nd edn. McGraw-Hill Companies, New York
90. Testa F, Iraldo F (2010) Shadows and lights of GSCM (green supply chain management): determinants and effects of these practices based on a multinational study. J Clean Prod 18:953–962
91. Tian G, Zhang H, Feng Y, Wang D, Peng Y, Jia H (2018) Green decoration materials selection under interior environment characteristics: a grey-correlation based hybrid MCDM method. Renew Sustain Energy Rev 81:682–692
92. Toloo M, Salahi M (2018) A powerful discriminative approach for selecting the most efficient unit in DEA. Comput Ind Eng 115:269–277
93. Tseng SC, Hung SW (2014) A strategic decision-making model considering the social costs of carbon dioxide emissions for sustainable supply chain management. J Environ Manage 133:315–322
94. UNEP (United Nations Environment Program) (2006) Environmental agreements and cleaner production: questions and answers. http://www.unep.fr/shared/publications/pdf/DTIx0833xPA-EnvAgreementsEN.pdf
95. Vahdani B, Mousavi SM, Tavakkoli-Moghaddam R, Hashemi H (2013) A new design of the elimination and choice translating reality method for multi-criteria group decision-making in an intuitionistic fuzzy environment. Appl Math Modell 37:1781–1799
96. Velasquez M, Hester PT (2013) An analysis of multi-criteria decision making methods. Int J Oper Res 10:56–66
97. Venkatachalam A, Anurag PVS, Sadanand TD, Nachimuthu R (2018) Optimization of the milling parameters for an Al/Si_3N_4 functionally graded composite using grey relational analysis. Mater Testing 60(2):215–221
98. Wang F, Lai X, Shi N (2011) A multi-objective optimization for green supply chain network design. Decis Support Syst 51:262–269
99. WCED (World Commission on Environment and Development) (1987) Report of the world commission on environment and development: our common future. Oxford University Press
100. Wilson FA (1997) The truth is out there: the search for emancipatory principles information systems design. Inf Technol People 10(3):187–204
101. Wu C, Barnes D (2016) An integrated model for green partner selection and supply chain construction. J Clean Prod 112:2114–2132
102. Xie G (2015) Modeling decision processes of a green supply chain with regulation on energy saving level. Comput Oper Res 54:266–273
103. Yang G, Wang Z, Li X (2009) The optimization of the closed-loop supply chain network. Transp Res E-Log 45:16–28
104. Yazdani M, Payam AF (2015) A comparative study on material selection of microelectromechanical systems electrostatic actuators using Ashby, VIKOR and TOPSIS. Mater Des 65:328–334
105. Yu C, Shao Y, Wang K, Zhang L (2019) A group decision making sustainable supplier selection approach using extended TOPSIS under interval-valued Pythagorean fuzzy environment. Expert Syst Appl 121:1–17
106. Yurdakul M, İç YT (2019) Comparison of fuzzy and crisp versions of an AHP and TOPSIS model for nontraditional manufacturing process ranking decision. J Adv Manuf Syst 18:167–192
107. Zhang K, Zhan J, Yao Y (2019) TOPSIS method based on a fuzzy covering approximation space: an application to biological nano-materials selection. Inf Sci 502:297–329
108. Zhang N, Wei G (2013) Extension of VIKOR method for decision making problem based on hesitant fuzzy set. Appl Math Model 37:4938–4947

109. Zhao R, Huang Y, Yu Y, Guo S (2019) An IVTIFN–TOPSIS based computational approach for pipe materials selection. Appl Sci Basel 9(24):5457
110. Zhao R, Liu S, Liu Y, Zhang L, Li Y (2018) A safety vulnerability assessment for chemical enterprises: a hybrid of a data envelopment analysis and fuzzy decision-making. J Loss Prev Process Ind 56:95–103
111. Zhao R, Neighbour G, Han J, McGuire M, Deutz P (2012) Using game theory to describe strategy selection for environmental risk and carbon emissions reduction in the green supply chain. J Loss Prev Process Ind 25(6):927–936
112. Zhao R, Zhong S (2015) Carbon labelling influences on consumers' behaviour: a system dynamics approach. Ecol Ind 51:98–106
113. Zhu QH, Dou YJ (2007) Evolutionary game model between governments and core enterprises in greening supply chains. Syst Eng Theory Pract 27:85–89
114. Zhu Q, Cote RP (2004) Integrating green supply chain management into an embryonic eco-industrial development: a case study of the Guitang Group. J Clean Prod 12:1025–1035
115. Zhu Q, Sarkis J (2004) Relationships between operational practices and performance among early adopters of green supply chain management practices in Chinese manufacturing enterprises. J Oper Manag 22:265–289
116. Zhu Q, Sarkis J, Geng Y (2005) Green supply chain management in China: pressures, practices and performance. Int J Oper Prod Manag 25:449–468
117. Zhu Q, Sarkis J, Lai KH (2007) Green supply chain management: pressures, practices and performance within the Chinese automobile industry. J Clean Prod 15:1041–1052
118. Zhu Q, Sarkis J, Lai KH (2008) Confirmation of a measurement model for green supply chain management practices implementation. Int J Prod Econ 111:261–273

Chapter 2
Application of MCDM to Improving Eco-efficiency of Supply Chain Management

2.1 Introduction

The increasing demand for consumption has greatly accelerated the depletion of resources and created environmental problems. Supply chain is no exception. Faced with the balance between profits and the environmental performance, the new consideration arouses the sustainable management for supply chain, not only to focus on waste reduction, energy savings, but also produce environmental benefits, thus improving eco-efficiency in terms of supply chain network [39].

Most of the existing studies developed optimization models based on maximizing eco-efficiency while minimizing environmental impacts on the supply chain network [22, 63, 84]. However, most manufactured materials in the supply chain would be considered hazardous to some extent if they are mismanaged or improperly disposed [95]. This fact is rarely considered in the literature. While economic and environmental performance are important, the safety and potential risks posed by the supply chain network are also important [53]. One of the major environmental concerns is the detoxification of industrial pollutants (Wang 2009), as many common industrial materials used in manufactured products can be considered harmful or 'hazardous' to the environment to a greater or lesser extent. In addition to the immediate risk posed by 'hazardous' materials in use, it is also important in the socio-political context to discriminate the "carbon footprint" within the supply chain, which may be an increasingly used indicator of the public acceptance of the product. Particularly, the environmental risk and the carbon emissions in the final disposal stage of a product are often very high compared to other stages [45, 105]. Decarbonisation thus has become a significant challenge for supply chain management, as it requires consideration of life cycle stages beyond the supply chain and the consumer. However, because the general framework of supply chain management is not usually considered (as opposed to the risk assessment of individual operations), manufacturers typically focus on the economic risks and benefits, as well as the supply

R. Zhao, *Multi-criteria Decision-Making Approaches to Sustainable Consumption and Production*, Urban Sustainability, https://doi.org/10.1007/978-981-96-2775-2_2

chain efficiency. The treatment and disposal of hazardous materials is usually given little deliberation especially in the post-consumer stage [18].

This Chapter considers how to re-design the conventional supply chain network of 'hazardous' materials in a more sustainable way and thus build MCDM models to minimize not only the total economic cost, but also the inherent risk and carbon footprint throughout the supply chain. In particular, followed by extending the responsibility of the supply chain, the inherent risk and carbon footprint in various scenarios, such as production, transportation, and disposal are taken into account for the model construction. Specifically, Sect. 2.2 proposes a multi-objective optimization model that promotes the minimization of supply chain environmental risk resulting from the use of hazardous materials, carbon emissions and economic costs. Section 2.3 takes the agro-product supply chain as a typical case to redesign it in order to improve its eco-efficiency by considering the associated agro-waste flowing into bioenergy enterprises for energy production. Since a variety of chemicals that can be considered harmful or hazardous are used in the supply chain network, chemical enterprises are vulnerable to safety accidents, such as fires and explosions, which pose potential hazards to human health and the environment [44, 87]. In section 2.4, a composite indicator system is proposed to assess the safety vulnerability of chemical enterprises by considering their exposure, sensitivity, and adaptability. The degrees of safety vulnerability are identified to provide insights into safety risk control and management for the key enterprise in the chemical supply chain network.

2.2 Optimization on Green Supply Chain Management

This Section presents a multi-objective optimization model for a green supply chain management scheme that minimizes the inherent risk caused by handling hazardous materials, the associated carbon emission and the economic cost. The model parameters are capitalized on data mining. Moreover, the raw data related to the parameters are derived from multiple sources, such as the enterprises' financial statements to obtain the data related to the business operation, empirical emission factors from the Intergovernmental Panel on Climate Change (IPCC) database to measure the carbon emissions, accident probability from the internal environmental information system (EIS) to measure the environmental impacts. To ensure the quality of the data, a statistical process control (SPC) chart is used to remove abnormal data from the dataset and extract the most representative data to be applied as the parameters of the model [29].

Since the green supply chain management attempts to increase economic returns by reducing environmental impacts, the extent to which economic costs, environmental risks, and carbon emissions reduction should be incorporated into management practices has been rarely discussed [93, 96]. The study aims to fill this gap by proposing an optimization model to optimize the safety, environmental and economic performance of the supply chain operations using the scenario analysis. The first scenario divides this optimization into three stages that first minimize environmental risk, and then carbon emissions and thus economic cost. The second minimizes both environmental risk and carbon emissions first, and then seeks to minimize economic cost. The last minimizes environmental risk, carbon emissions and economic cost simultaneously. Additionally, most of the previous studies mainly focused on the optimization process to obtain optimal values of variables corresponding to the predefined objectives. In this study, the MODM and MADM approaches are combined to solve the model and rank the priority of each scenario, in which the optimal scenario is determined to help improve the management practice of green supply chain. It is expected that the multi-optimization model can improve risk control and carbon emissions reduction with regard to supply chain management, and also inform the decision-making of involved stakeholders on supply chain design, thus promoting sustainable development.

2.2.1 Model Construction

The proposed supply chain network consists of a number of raw materials suppliers, a key manufacturing enterprise, and several distribution centers. This proposed network considers common disposal centers (landfill, incineration plant and recycling centers) in the illustration of a full life cycle, as shown in Fig. 2.1. To simplify the supply chain, the following hypothesis are proposed:

(1) The locations of the raw material suppliers, key manufacturer and distribution centers are given in advance, as well as their capacities, and transportation routes are identified.

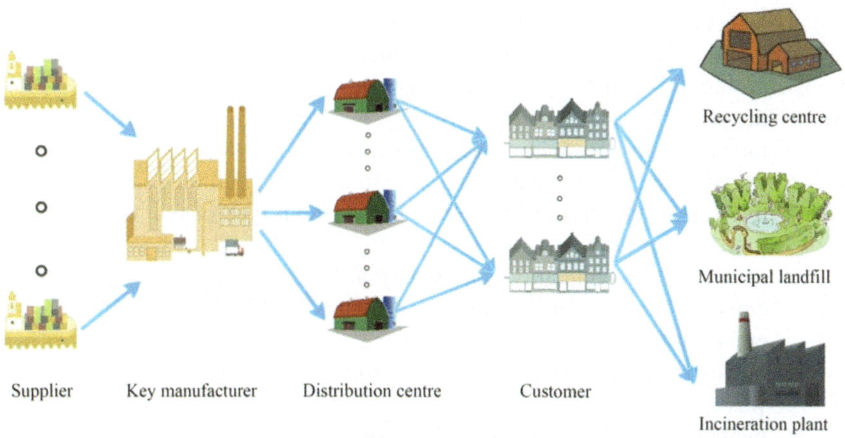

Fig. 2.1 The proposed green supply chain network

(2) The supply chain is deemed to be a single-period product supply chain, which assumes that facilities do not keep inventory. Regardless of the type of product, one unit of production capacity is required to produce one unit of product.
(3) There is no difference between remanufactured products and manufactured products.
(4) The suppliers are considered as an entity without differences, e.g. the possible market competition, variance of product quality, the possible coordination, etc.

The corresponding parameters of the proposed supply chain network are given as follows:

s set of suppliers
k key manufacturer
d set of distribution centres
c set of potential consumers
r recycling centre
I incineration plant
la municipal landfill

The input parameters to cover the above indices include the following:

dem_k raw materials demand of key manufacturer, kg
$demc$ demand of consumer c, kg
cr_s cost of purchasing raw materials from supplier s
cff' unit transportation cost from facility f to facility f′, f, f′∈{s,k,d,c,r,i,la}
fc_f fixed cost of operating facility, f ∈{s,d,r,i,la}
cp_f processing cost per unit product in facility, f ∈{s,d,r,i,la}
ca_f capacity of facility, f ∈{s,d,r,i,la}
ca_t capacity of transportation
inv_m investment for risk control when the risk level is m
inv_l investment for carbon emissions reduction when the emission level is l
$Prob_m$ probability of risk when the level is m
$Prob_f$ the probability of risk in facility f
$Prob_{ff'}$ the probability of transportation risk from f to f′
N_{fp} the consequence of the p type of risk in facility f, million Yuan
$N_{ff'p}$ the transportation accident resulted from the p type of risk from f to f′, p ∈{1,2,3}, million Yuan
L_{ij} distance between facility f ∈{s,k,d,c} and f′ ∈{k,d,c,r,i,la}, km
EM_l emission factor when the emission level is l, kg CO_2/kg
EM_f emission factor of facility f, kg CO_2/kg
$EM_{ff'}$ transportation emission factor from f to f′, kg CO_2/kg
η waste generation rate
ξ incineration ratio of unrecyclable waste
ω landfilling ratio

The decision variables related to the above indices are given as follows:

$$x_s = \begin{cases} 1, \text{ if supplier s is selected} \\ 0, \text{ otherwise} \end{cases}$$

$$x_d = \begin{cases} 1, \text{ if distribution center d is selected} \\ 0, \text{ otherwise} \end{cases}$$

$$x_r = \begin{cases} 1, \text{ if recycling centre r is operated} \\ 0, \text{ otherwise} \end{cases}$$

$$x_I = \begin{cases} 1, \text{ if incineration plant is operated} \\ 0, \text{ otherwise} \end{cases}$$

$$x_{la} = \begin{cases} 1, \text{ if landfill la is operated} \\ 0, \text{ otherwise} \end{cases}$$

$$y_m = \begin{cases} 1, \text{ if risk level m is selected when taking risk control measures} \\ 0, \text{ otherwise} \end{cases}$$

$$z_l = \begin{cases} 1, \text{ if carbon emissions level l is selected when taking emissions reduction measures} \\ 0, \text{ otherwise} \end{cases}$$

μ the recycling rate of recycling centre
x_{sk} the amount of products flowing from supplier s to the key manufacturer
x_{kd} the amount of products flowing from the key manufacturer to distribution centre d
x_{dc} the amount of products flowing from distribution centre d to customer c
x_{cr} the amount of products flowing from customer c to recycling centre r
x_{ci} the amount of products flowing from customer c to incineration plant I
x_{cla} the amount of products flowing from customer c to municipal landfill la

The optimization model is complied with three objectives, i.e., minimization of inherent risk by 'hazardous materials', carbon emissions, and total economic cost, which are presented as follows.

Minimizing inherent risk

The inherent risk of 'hazardous materials' supply chain is comprised of three sub-risks, i.e. risk of casualties, risk of environmental pollution and risk of property loss, which is deemed as the possible consequence multiplied by the associated probability in a hazard scenario [53, 62, 95]. More specifically, the risk of casualties, environmental pollution and property loss are defined as the casualties, water, air and soil pollution, and direct economic loss, respectively, caused by an emergency accident in the process of supply chain operation, given as follows [55, 66]:

$$R_i = Prob_i \cdot \sum_{p=1}^{3} N_{ip} \tag{2.1}$$

where R_I indicates the risk in facility I, $Prob_I$ the probability of the failure events in facility I, N_{ip} the consequence of risk type p, including the casualties, property loss and environmental pollution.

Transportation risk during supply chain management can be expressed as follows [71]:

$$R_{ij} = Prob_{ij} \cdot \sum_{p=1}^{3} Nijp \tag{2.2}$$

$$Prob_{ij} = Ar_{ij} \cdot L_{ij} \cdot Q_{ij} \tag{2.3}$$

where $Prob_{ij}$ represents the probability of risk from facility I to j, L_{ij} the path-length, Q_{ij} the amount of transported products, Ar_{ij} the accident rate of vehicle transport.

The objective function of minimizing inherent risk is given as follows:

$$OBJ1 = min(R_f + R_t) \tag{2.4}$$

$$\begin{aligned} R_f = & \sum_s Prob_s \sum_{p=1}^{3} N_{sp} + \sum_m Prob_m \cdot y_m \cdot \sum_{p=1}^{3} N_{kp} \\ & + \sum_d Prob_d \sum_{p=1}^{3} Ndp + \sum_c Probc \sum_{p=1}^{3} Ncp + Prob_r \cdot \sum_{p=1}^{3} N_{rp} \\ & + Prob_I \cdot \sum_{p=1}^{3} N_{ip} + Prob_{la} \cdot \sum_{p=1}^{3} N_{lap} \end{aligned} \tag{2.5}$$

$$\begin{aligned} R_t = & \sum_s x_{sk} \cdot Ar_{sk} \cdot L_{sk} \sum_{p=1}^{3} N_{skp} + \sum_d x_{kd} \cdot Ar_{kd} \cdot L_{kd} \sum_{p=1}^{3} N_{kdp} \\ & + \sum_d \sum_c x_{dc} \cdot Ar_{dc} \cdot L_{dc} \sum_{p=1}^{3} N_{dcp} + \sum_c x_{cr} \cdot Ar_{cr} \cdot L_{cr} \sum_{p=1}^{3} N_{crp} \\ & + \sum_c x_{ci} \cdot Ar_{ci} \cdot L_{ci} \sum_{p=1}^{3} N_{cip} + \sum_c x_{cla} \cdot Ar_{cla} \cdot L_{cla} \sum_{p=1}^{3} N_{cla} \end{aligned} \tag{2.6}$$

where R_f measures the risk in all the facilities, from the upstream suppliers to the disposal centre, whilst R_t measures the transportation risk based on the flow of products from different nodes.

Minimizing carbon emissions

The total carbon emissions of the supply chain can be calculated as follows:

$$CE = \sum_{i} Q_i \cdot EM_I \tag{2.7}$$

where Q_I indicates the ith activity contributing carbon emissions, e.g. production, transportation, disposal etc., EM_I the emission factor of the ith activity.

Thus, the objective function of minimizing carbon emission is given as follows:

$$OBJ2 = min(CE_f + CE_t) \tag{2.8}$$

where:

$$\begin{aligned} CE_f = {} & \sum_{s} x_{sk} \cdot EM_s + \sum_{l} EM_l \cdot zl \cdot demk + \sum_{d}\sum_{c} x_{dc} \cdot EM_c \\ & + \sum_{c} x_{cr} \cdot EM_r + \sum_{c} x_{ci} \cdot EM_I + \sum_{c} x_{cla} \cdot EM_{la} \end{aligned} \tag{2.9}$$

$$\begin{aligned} CE_t = {} & \sum_{s} x_{sk} \cdot EM_{sk} \cdot L_{sk} + \sum_{d} x_{kd} \cdot EM_{kd} \cdot L_{kd} + \sum_{d}\sum_{c} x_{dc} \cdot EM_{dc} \cdot L_{dc} \\ & + \sum_{c} x_{cr} \cdot EM_{cr} \cdot L_{cr} + \sum_{c} x_{ci} \cdot EM_{ci} \cdot L_{ci} + \sum_{c} x_{cla} \cdot EM_{cla} \cdot L_{cla} \end{aligned} \tag{2.10}$$

where CE_f denotes the carbon emissions in all the facilities on the supply chain, CE_t the carbon emissions of transportation, based on the transportation distances.

Minimizing total economic cost

In this study, the total cost of the supply chain not only includes the cost of production, transportation and waste disposal, but also the costs paid to reduce the inherent risk and carbon emissions, shown as follows:

$$OBJ3 = \min\left(FC + RC + MC + DC + TC + \sum_{m} Inv_m \cdot ym + \sum_{l} Inv_l \cdot zl\right) \tag{2.11}$$

where

$$FC = \sum_{s} fc_s.x_s + \sum_{d} fc_d.x_d + fc_r.x_r + fc_i.x_i + fc_{la}.x_{la} \tag{2.12}$$

$$RC = \sum_{s} x_{sk}.cr_s \tag{2.13}$$

$$MC = \sum_{s} x_{sk}.cp_{s} + \sum_{d} x_{kd}.cp_{d} \tag{2.14}$$

$$DC = \sum_{c} x_{cr}.cp_{r} + \sum_{c} x_{ci}.cp_{i} + \sum_{c} x_{cla}.cp_{la} \tag{2.15}$$

$$TC = \sum_{s} x_{sk}.c_{sk}.L_{sk} + \sum_{d} x_{kd}.c_{kd}.L_{kd} + \sum_{d}\sum_{c} x_{dc}.c_{dc}.L_{dc} + \sum_{c} x_{cr}.c_{cr}.L_{cr} + \sum_{c} x_{ci}.c_{ci}.L_{ci} + \sum_{c} x_{cla}.c_{cla}.L_{cla} \tag{2.16}$$

where FC indicates the fixed cost based on operation of the involved facilities, RC the purchasing cost of raw materials, MC the cost of raw materials extraction and products manufacturing, DC the cost of waste disposal, TC the transportation cost, the $\sum_{m} Inv_m \cdot y_m$ the investment of risk control and $\sum_{l} Inv_l \cdot z_l$ the investment of carbon emissions reduction.

Some generic constraints are considered in the optimization model based on the proposed decision variables, including balance constraints of materials, the capacity limit constraint, and the constraints of decision variables, which are presented below.

Mass balance constraints

$$\sum_{s} x_{sk} = demk \tag{2.17}$$

$$\sum_{d} x_{kd} = \sum_{c} dem_{c} \tag{2.18}$$

$$x_{kd} = \sum_{c} x_{dc} \forall d \tag{2.19}$$

$$dem_{c} = \sum_{c} x_{dc} \forall c \tag{2.20}$$

$$x_{cr} = dem_{c} \times \eta \times \mu \forall c \tag{2.21}$$

$$x_{ci} = dem_{c} \times \eta \times (1 - \mu) \times \xi \forall c \tag{2.22}$$

$$x_{cla} = dem_{c} \times \eta \times (1 - \mu) \times \omega \forall c \tag{2.23}$$

where constraint (2.17) ensures the demand of raw materials for the key manufacturer, and constraints from (2.18) to (2.20) define that the products provided by the key manufacturer are equal to the demand of potential consumers, and the products flowing from the key manufacturer are equal to the distribution centres received. Constraints (2.21) to (2.23) assure the mass balances of waste generated and recycled.

Capacity limit constraints

$$x_{sk} \leq ca_s \cdot x_s \forall s \tag{2.24}$$

$$\sum_d x_{kd} \leq ca_k \tag{2.25}$$

$$\sum_d x_{dc} \leq ca_d \cdot x_d \forall d \tag{2.26}$$

$$\sum_c dem_c \cdot \eta \cdot \mu \leq ca_r \cdot x_r \tag{2.27}$$

$$x_{ff'} \leq ca_{ff'} \tag{2.28}$$

where constraints from (2.24) to (2.27) ensure the capacity limits in each facility, which are less than the designed capacity, whilst constraint (2.28) ensures that the transported products cannot exceed the transportation capacity.

Constraints of decision variables

$$x_s \leq x_{sk} \forall s \tag{2.29}$$

$$x_d \leq \sum_c x_{dc} \forall d \tag{2.30}$$

$$x_r \leq \sum_c dem_c \cdot \eta \cdot \mu \tag{2.31}$$

$$x_s, x_d, x_r, x_i, x_{la} \in \{0, 1\} \tag{2.32}$$

$$x_{ff'} \geq 0 \tag{2.33}$$

$$y_m, z_l \in \{0, 1\} \text{and}\, m \in \{1, \ldots, a\}, l \in \{1, \ldots, b\}, a, b \in Z \tag{2.34}$$

$$\mu \in [0.1] \tag{2.35}$$

where constraints from (2.29) to (2.31) define the lower production capacity limits, constraints from (2.32) to (2.33) define the decision variables $x_s, x_d, x_r, x_i, x_{la}$ are binary integer variables, which are non-negative. Constraint (2.34) definesy_m, Z_l are binary integer variables, where m and 1 are integers within the interval [1, a], [1, b] respectively, which represents the different level of risk and carbon emissions reduction while taking the corresponding control measures. Constraint (2.35) defines that the recycling rate μ is within the interval [0, 1].

2.2.2 An Illustrative Case

A local supply chain network of a sanitary product in Sichuan Province, China, is taken as an illustrative case example to verify the proposed optimization model, as shown in Fig. 2.2a. There is a key manufacturer (A) on the supply chain network, located in Jianyang City, which is the largest enterprise in Sichuan Province that produces sanitary ware using acrylic materials (e.g. water closets, bathtubs, shower stalls, etc.) Three potential suppliers are located in Jianyang City, Longquanyi District and Ziyang City, which are indicated in Fig. 2.2a as B, C, D, respectively. These suppliers can provide acrylic board to A and other sanitary ware factories in Sichuan province. The sanitary ware products from the key manufacturer A are transported to the distribution centers located in Chongqing and Chengdu, denoted as E and F, respectively. After the products are used, waste disposal is considered, some of which is recycled in the recycling center of manufacturer A and used for remanufacturing. Other waste that cannot be recycled is disposed of in municipal landfills or incinerated in Chongqing and Chengdu. The geographic location of the local supply chain network is shown in Fig. 2.2b.

The data related to the parameters of economic cost were obtained from the field investigation, which focused mainly on the financial statements of the enterprises involved in the supply chain network. The raw data of various costs for the period from 2010 to 2014 were collected from each facility, and the scatter diagrams were generated by using SPSS 19.0 to investigate the upper control limit (UCL) and lower control limit (LCL). The data that deviated from the UCL and LCL were eliminated, and the mean values were input into the model. In order to show how to use the data analytic approach to extract the meaningful data, the demand of raw materials from the key manufacturer is taken as an example to demonstrate the data processing by using a schematic diagram, as shown in Fig. 2.3. The data from the financial statement of the key manufacturer from the year 2010 to 2014 was imported into the SPSS 19.0 to generate a control chart. The UCL and LCL were set as $\pm$ 3 times of the standard deviation according to the hypothesis test [29, 56]. The data that deviated from the LCL, shown in Fig. 2.3, are deemed as the abnormal data that will be removed. Moreover, the mean of the dataset, such as 3652 kg/d, was determined as the representative data to be input into the optimization model as a parameter. Similarly, other parameters can be derived in the same way, as shown in Table 2.1.

Carbon emissions in the stages of materials supply, manufacture and transportation are converted by energy consumption, e.g. electricity, coal, and diesel. Data on energy consumption was obtained from the audit report of cleaner production of the investigated enterprises in the supply chain. In particular, data on the different energy consumption of the key manufacturer were obtained from the internal report of energy saving and emissions reduction.

For waste disposal, carbon emissions from landfill disposal are converted to methane emissions, given as follows [36]:

$$\mathrm{CH_4}\ emission = A \times k \times MSW_T \times MSW_F \times L_0 \tag{2.36}$$

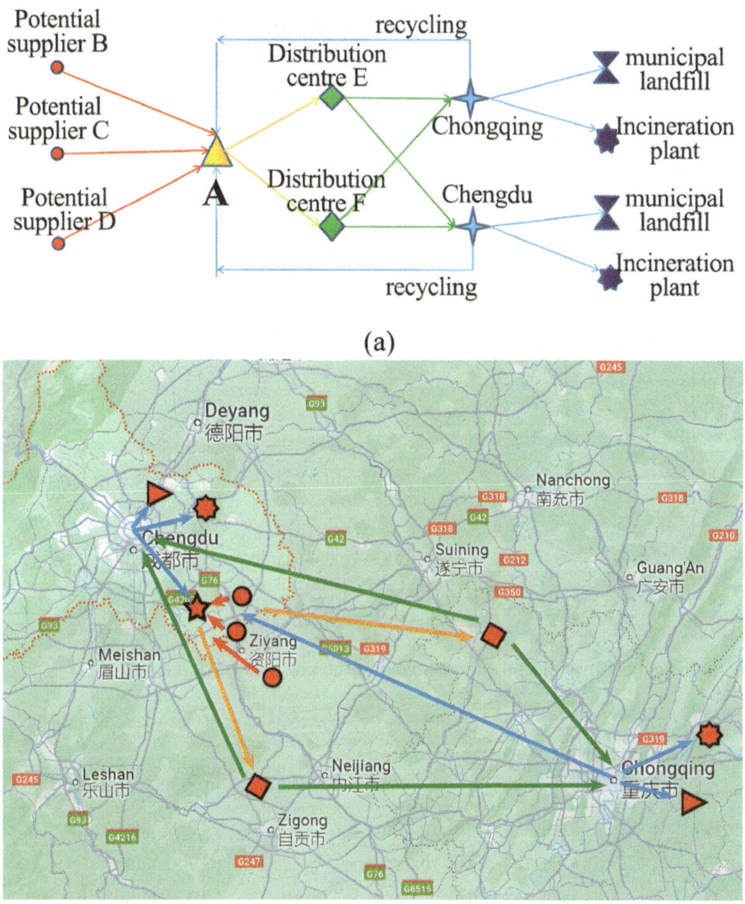

Fig. 2.2 The proposed green supply chain network in the case example

$$L_0 = MCF \times DOC \times DOC_F \times F \times 16/12 \tag{2.37}$$

where A indicates the normalization factor, k is the constant of the methane generation rate, MSW_T is the amount of municipal waste generated, MSW_F is the fraction of landfill disposal of municipal waste, L_0 is the methane generation potential, MCF is the methane correction factor, DOC is the degradable organic carbon, DOC_F is the fraction of the degradable organic carbon dissimilated, and F is the fraction of methane by volume the occurs in landfill gas.

The CO_2 emission of the incineration can be calculated by [36]:

$$CO_2\,emission = \sum_i SWi \times dmi \times CFi \times FCFi \times OFi \times 44/12 \tag{2.38}$$

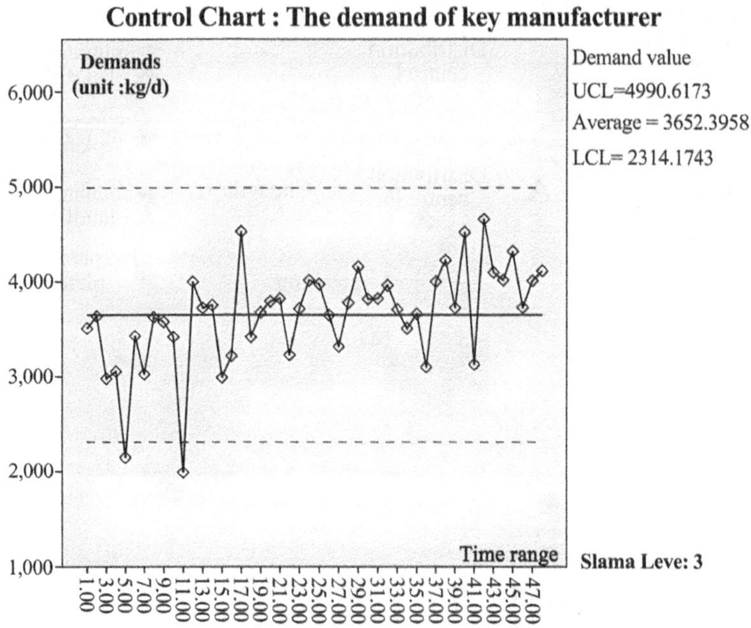

Fig. 2.3 Statistical control chart for the demand of the key manufacturer

Table 2.1 Measurement of the economic cost

Facility	FC	MC	DC	RC	TC
Supplier B	12,330 Yuan/d	5500 Yuan/t	×	1960 Yuan/t	×
Supplier C	12,300 Yuan/d	5000 Yuan/t	×	1900 Yuan/t	×
Supplier D	12,300 Yuan/d	5000 Yuan/t	×	1900 Yuan/t	×
Distribution centre E	950 Yuan/d	300 Yuan/t	×	×	×
Distribution centre F	1000 Yuan/d	300 Yuan/t	×	×	×
Recycling centre	×	×	– 100 Yuan/t[a]	×	×
Landfill	×	×	50 Yuan/t	×	×
Incineration plant	×	×	100 Yuan/t	×	×
Transportation	×	×	×	×	10 Yuan/t km

where SW_i represents amount of the ith type of waste incinerated or open-burned, dmi is the dry matter content in the waste that is incinerated or open-burned, CF_i is the fraction of carbon in the dry matter, FCF_i is the fraction of fossil carbon in the total carbon, and OF_i is the oxidation factor.

All empirical values related to the disposal stage (DOC, DOC_F, CF_i, FCF_i and OF_i) are derived from the IPCC guidelines, while others are derived from the work of Yang [90] and Liu et al. [50]. The carbon emissions for each facility are shown in Table 2.2.

Risk is quantified by multiplying the hazard probability by the possible consequences [5, 95]. The acceptable risk level is used to measure the hazard probability [33]. To understand what the level of risk should be, the principle of "Tolerability of Risk" provides a benchmark to determine the level of acceptability [32, 96]. Risk is deliberately divided into three levels: intolerable, tolerable and acceptable, as shown in Fig. 2.4.

Analogous to the HSE guideline, the maximum risk (i.e. the accident rate in any facility in the supply chain network) to any member of the public that should be tolerated is set at 1 in 10^4 per year. For transportation, the tolerable traffic accident rate is set at 2.4×10^{-4} (frequent accidents), and 1.61×10^{-5} (typical number of accidents), respectively [33].

Table 2.2 Measurement of the unit carbon emissions

Facility	Unit carbon emissions	Unit
Supplier B	623.15	kg CO_2 per tonne of product
Supplier C	752.04	kg CO_2 per tonne of product
Supplier D	785.82	kg CO_2 per tonne of product
Key manufacturer A	268.65 (low cost plan)	kg CO_2 per tonne of product
	247.64 (medium cost plan)	
	246.21 (high cost plan)	
Recycling center	523.51	kg CO_2 per tonne of waste
Landfill	970.2	kg CO_2 per tonne of waste
Incineration plant	240	kg CO_2 per tonne of waste
Transportation	0.14	kg CO_2 per tonne kilometers

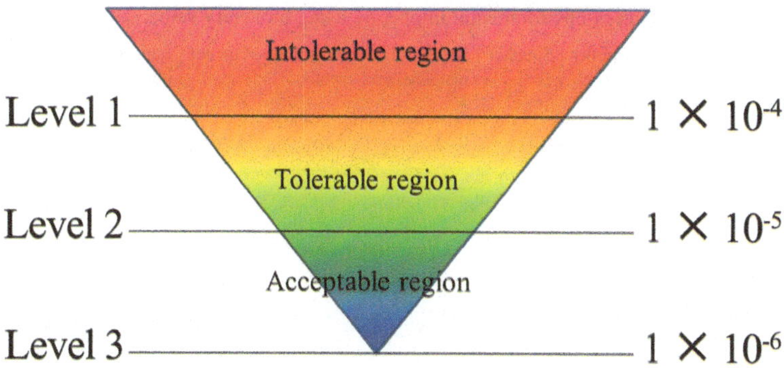

Fig. 2.4 The adjusted framework of risk tolerability

The consequences of risk are mainly divided into three categories: casualties, environmental pollution and property loss. According to the field investigation, the major risk may happen in the stage of raw materials supply and waste disposal. Suppliers B, C and D may experience explosion accidents in the process of the preparation of acrylic sheets due to the volatility and flammability of the raw materials as Methl-Methacrylate (MMA). For waste disposal, the risk comes from landfilling and incineration, and mostly involves substances such as waste leachate and landfill gas [97].

The 'Regulation of Production Safety Accident Report and Investigation of China' [80] was implemented to quantify casualties and property losses that occur at facilities in the supply chain network. This document was calibrated by statistical analysis of the risks inherent in the investigated enterprise. The corresponding results are shown in Table 2.3.

To quantify the environmental impact, the authors propose to multiply the impacted areas by the cost of emergency treatment, such as emergency rescue, monitoring, and remediation of an impacted area, that various accidents require. For example, a water pollution accident required adsorption treatment with activated carbon, and the unit cost was 0.04 Yuan per cubic meter in terms of the dosage of activated carbon. In the disposal stage, the environmental impact of a municipal landfill and incineration plant can be quantified by the cost of in-situ remediation required when waste leachate infiltrates the groundwater. The environmental impact of the supply chain network using a monetary measure is shown in Table 2.4.

Table 2.3 Measurement of possible casualties and property loss using monetary value

	Supplier B	Supplier C	Supplier D	Key manufacturer A	Recycling center	Landfill	Incineration plant	Transportation
Consequence (billion Yuan)	5.50	4.89	4.89	10.00	3.00	0.20	1.00	1.50

Table 2.4 Measurement of environmental impact by using monetary value

	Supplier B	Supplier C	Supplier D	Key manufacturer A	Recycling center	Landfill	Incineration plant	Transportation
Consequence (billion Yuan)	5.00	0.91	0.24	0.55	3.16	10.00	9.00	0.60

2.2.3 *Optimization Results*

The model aims at improvement of the existing risk control and reduction of carbon emissions to promote the sustainable development of the green supply chain management. Three scenarios are proposed to solve the optimization model. In the first scenario, the optimization seeks to minimize the risk first, and then the carbon emission and thus the economic cost. The second scenario seeks to minimize both risk and carbon emission first, and then seeks to minimize the economic cost, while the third scenario attempts to minimize the risk, carbon emission and economic cost simultaneously.

Scenario one

In this scenario, the minimum solution of risk was taken as a constraint to obtain the minimum carbon emission, which was substituted into the objective function of total economic cost to calculate the minimum cost. In addition, Lingo programming is applied to solve the proposed model. The soulution process and the correspondisolutions are shown in Fig. 2.5, where the supplier D and distribution centre F are selected as the operating facilities. The recycling rate of the recycling centre is set above 50%. The level of risk control and carbon emissions reduction is set as the third level, indicating that the highest standard is implemented by the stakeholders involved. The numerical flow of raw materials and products in the supply chain network is shown with the arrows.

Scenario two

In the second scenario, the ε-constraints method was used to obtain the lower bound ε for risk and carbon emissions, which were substituted into the objective function of economic cost as the constraint conditions used to seek the optimal solution. The process of the solution and the corresponding results are shown in Fig. 2.6.

Scenario three

In the third scenario, the 'ideal point method' was proposed to solve the optimization model. Here, the proximity used to measure the difference between the ith objective function and its ideal point is defined as follows:

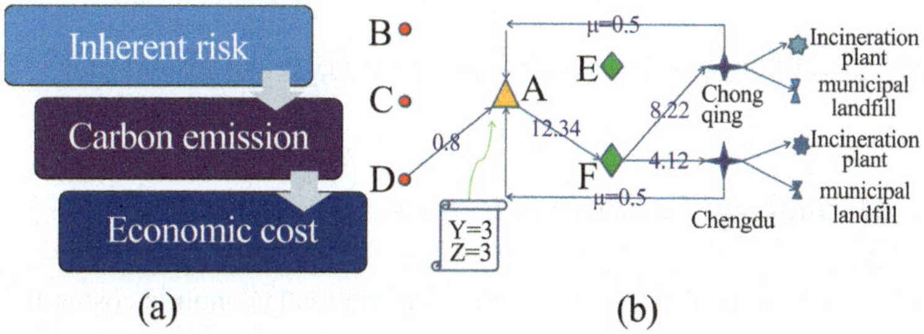

Fig. 2.5 Solution of scenario 1

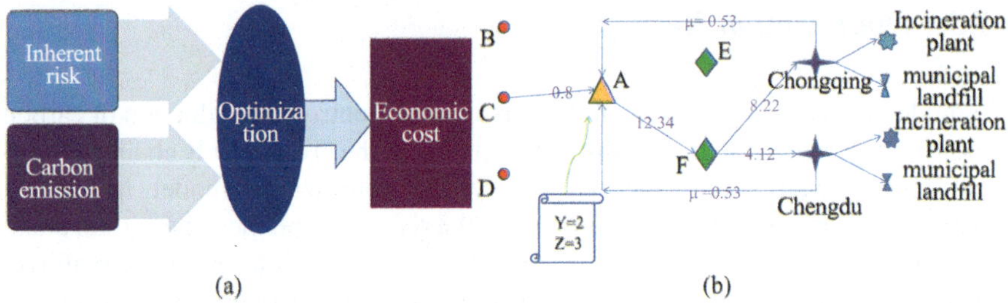

Fig. 2.6 Solution of scenario 2

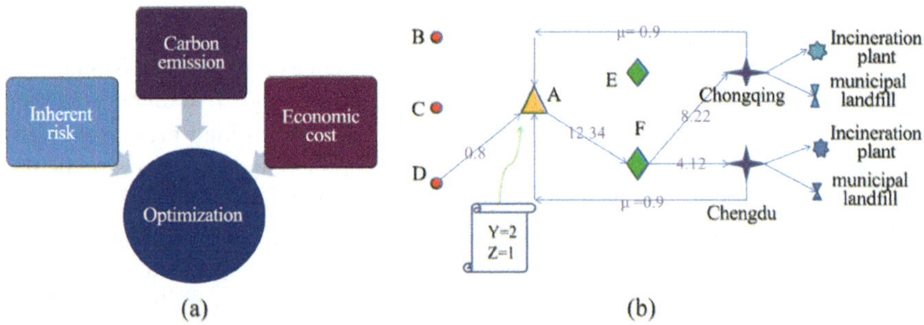

Fig. 2.7 Solution of scenario 3

$$d_i = \frac{OBJ_i}{OBJ_{i*}} \tag{2.39}$$

where OBJ_i represents the Pareto-solution of the ith objective, and OBJ_{i*} is the optimal solution under the constraint of a single objective function, indicated as the 'ideal point'.

If the d_i is equal to 1, then the optimal solution of the objective function is the ideal solution. Thus, multi-objective functions are transformed into a single objective function as below:

$$minf = \frac{OBJ_{cost}}{OBJ_{cost*}} + \frac{OBT_{CF}}{OBJ_{CF*}} + \frac{OBJ_{risk}}{OBJ_{risk*}} \tag{2.40}$$

The process of the solution and the corresponding results are shown in Fig. 2.7.

2.2.4 Managerial Scenario Selection

The optimal solutions of risk, carbon emission and total economic cost in the three pre-defined scenarios are shown in Table 2.5, where it can be seen that each scenario has its own advantages and disadvantages in terms of its impact on the sustainability of the supply chain network. For instance, scenario one has the lowest risk but the highest total cost. In scenario two, the carbon emissions are the lowest, but the risk is the highest. It is difficult to decide which scenario will best help stakeholders in their efforts to improve supply chain management. A multi-attributive decision making (MADM) approach is proposed to fill this gap. This approach uses a performance assessment that multiplies the weight of each objective by the corresponding attributive value to discover the optimal scenario for future supply chain management practices.

The 'Binary Dominance Matrix' (BDM) is applied to assess the importance of the three objectives (risk, carbon emissions and economic cost). This is a subjective weighting approach that lists all the criteria on both the vertical and horizontal axes of a matrix, and assigns a value of 1, 0.5 or 0 to each box generated by the intersection of the axes, according to the relative importance of a pair of indicators relative to each other [86, 95]. To ensure the arithmetic validity of this approach, each row of the matrix should add up to 1 when all the indicators are scored. The scores and their weights corresponding to the three scenarios are shown in Table 2.6.

Table 2.5 Optimal objective results related to different scenarios

	Inherent risk (million Yuan)	Carbon emissions (kg CO_2)	Total cost (million Yuan)
Scenario 1	40.94	3540.01	1.17
Scenario 2	54.47	3499.54	0.47
Scenario 3	54.42	4087.68	0.27

Table 2.6 Weight determination of different scenarios using binary dominance matrix

Indicators	Inherent risk	Carbon emissions	Total Cost	Scores	Weights
(a) Scenario 1					
Inherent risk	×	1	1	3	0.5
Carbon emissions	0	×	1	2	0.33
Total Cost	0	0	×	1	0.17
(b) Scenario 2					
Inherent risk	×	0.5	1	2.5	0.42
Carbon emissions	0.5	×	1	2.5	0.42
Total Cost	0	0	×	1	0.17
(c) Scenario 3					
Inherent risk	×	0.5	0.5	2	0.33
Carbon emissions	0.5	×	0.5	2	0.33
Total Cost	0.5	0.5	×	2	0.33

Due to the different dimensions, normalization is conducted to ensure a common scale for comparison, given as follows:

$$A_{ij} = \frac{OBJ_{ij} - OBJ_{imin}}{OBJ_{imax} - OBJ_{imin}} \tag{2.41}$$

where A_{ij} indicates the normalized attributive value of objective i in scenario j, OBJ_{ij} is the actual value of objective i in scenario j, OBJ_{imin} is the minimum value of objective i among the three scenarios, and OBJ_{imax} is the maximum value of objective i among the three scenarios. The normalized attributive values of three scenarios are shown in Table 2.7.

The performance of the three scenarios can be ultimately assessed by multiplying the weights by the attribute value of each objective, as shown in Table 2.8. When all the objectives are set in this minimization function, scenario one, which intends to reduce the inherent risk at first, then the carbon emissions, and finally the total economic cost, appears superior to scenarios two and three.

However, the key stakeholders involved in the supply chain network (suppliers, manufacturers and distributers) may prefer to implement the strategies that reduce risk and carbon emissions (i.e. scenarios 2 or 3) rather than choose scenario 1, due to the comparatively high costs of the scenario 1. There are three levels of sustainable business from bottom to top, which can be described as basic market demand (bottom line), sustainable sources of raw materials and energy for production, the gradual achievement of "green" social value for environmental protection [4, 40, 77, 79, 85, 100]. The fundamental role of an organization involved in the supply chain is to promote the economic stability of its individual firm while acting in accordance with policies and behavioral norms [4, 77]. Even in green supply chain management, which considers potential savings or profits as a driving force, economic issues play a dominant role in business operations [19, 54, 98]. As the sense of corporate social responsibility is gradually emerging in the green supply chain management, business entities should take a longer-term perspective when evaluating their performance in terms of sustainability issues. Such a perspective would encourage a positive attitude among the involved stakeholders toward environmental improvement,

Table 2.7 Attributive values of each objective related to different scenarios

	Inherent risk	Carbon emissions	Total costs
Scenario 1	0	0.07	1
Scenario 2	1	0	0.22
Scenario 3	0.996	1	0

Table 2.8 Evaluation results of different scenarios

	Scenario 1	Scenario 2	Scenario 3
Evaluation results	0.19	0.46	0.67

especially the organizations involved will gradually seek to benefit from the business opportunities associated with the sales of 'environmentally friendly products' [40, 52, 65, 72].

Although the total economic cost is comparatively high in scenario 1, some of them is considered to be one-off investments in risk control and carbon emission reduction. Such up-front investment may promote green innovation and market competition and thus promoting long-term commercial success [13, 23, 85]. It is further expected that the sales revenue of perceived 'green' products may fully cover the cost of risk control and carbon emissions reduction [99, 100]. In addition to the internal driving force from the self-interest of enterprises, the external factors, i.e. government policy instruments are important to promote the green transitions of the supply chain [100, 104]. The policy instruments are usually divided into the incentive mechanism and punitive mechanism [59, 88, 96]. The incentive mechanism consists of subsidies, tax preference, price regulation, etc., which have been found to be effective in reducing the externalities of negative consequences during the implementation of the green strategy, decreasing the financial risk of related stakeholders, and encouraging them to take more corporate social responsibilities (CSR) to improve their social performance, ultimately seeking sustainability in commercial success ([,57, 70]). As for the punitive mechanism, such as economic sanction, legal supervision, etc., it is a compulsive measure to drive innovation, research and design for product sustainability, to reduce lifecycle environmental impact, thus improving the eco-efficiency of the supply chain network [8, 11, 25].

In order to help the involved enterprises to choose scenario 1 for their green transition, it is suggested that the government provides incentives to cover part of the total economic costs in the initial stage. When their inherent risk and carbon emissions have been decreased, it is further suggested that the government gradually imposes supervision to ensure the sustainability of the supply chain.

2.2.5 Summary

This section offers a multi-objective optimization model to provide insight into green supply chain management while minimizing the risks inherent in the handling of hazardous materials; such risks are generally associated with carbon emissions and economic costs. Three scenarios are proposed to solve the optimization model, where scenario 1 minimizes risk first, then carbon emissions, and finally economic cost. Scenario 2 seeks to minimize economic cost by minimizing both risk and carbon emissions, while scenario 3 attempts to minimize risk, carbon emissions and economic cost simultaneously. A case study of the sanitary ware supply chain network is given to demonstrate the application of the optimization model. This study shows that scenario 1 is superior to scenarios 2 and 3 in reducing risk and carbon emissions, promoting improvements in green supply chain management, and achieving long-term commercial success. It is expected that the best strategy of risk control and carbon emissions reduction can be determined through the application of the optimization, which is insightful to be incorporated into the management practice, thus improving green supply chain management.

2.3 Redesign of a Closed-Loop Supply Chain

The supply chains of agro-products usually have low eco-efficiency, which leads to lower utilization of agro-wastes and thus affects the formulation of reverse logistics [73]. As these agro-wastes usually contain large amounts of biomasses, recycling them for energy recovery can provide greater economic and environmental benefits [2, 31, 58]. However, the incorporation of bioenergy enterprises into the supply chain inevitably disrupts supply chain operations [51]. For instance, it may increase the number of stakeholders involved, making the coordination of the supply chain network more difficult [76]. Meanwhile, additional costs may be incurred, including the cost of agro-waste collection, pretreatment, and transportation, resulting in an impact on the value added to the supply chain [64, 81].

This Section redesigns the supply chain of agricultural products in Southwestern China to improve its eco-efficiency by considering the associated agro-waste flowing into a bioenergy enterprise for energy production. Two scenarios are created, in which the first assumes that all wastes flow into the enterprise, while the second considers only the inflow of agro-wastes produced by farmers and the wholesale market. A system dynamics simulation is conducted by using carbon emissions per product as an indicator to obtain the optimal scenario for management practice and to design an incentive mechanism to drive supply chain operations.

2.3.1 Model Formulation

A simplified agro-product supply chain includes farmers, market, consumers, and a municipal waste disposal center. Farmers are primarily responsible for agricultural production. The "market" indicates a retailer in a large wholesale market that buys agro-products at a market price and sells them to the consumers (i.e. urban residents). The agro-wastes from the farmers, retailers, and consumers are sent to the municipal landfill for final disposal. By including bioenergy enterprises in the supply chain network, agro-wastes can be recovered for energy production. Their energy production will be integrated into the local grid system to provide renewable energy consumed by urban residents (consumers) and farmers, reduce waste disposal in landfills, and promote energy savings.

To explore the optimal pattern of supply chain operations against this background, two scenarios are established in this study: in Supply Chain 1, all the agro-wastes produced by farmers, market, and consumers proceed to the bioenergy enterprise for processing as shown in Fig. 2.8a; in Supply Chain 2, agro-wastes produced by consumers are sent to the landfill, while the wastes produced by farmers and the

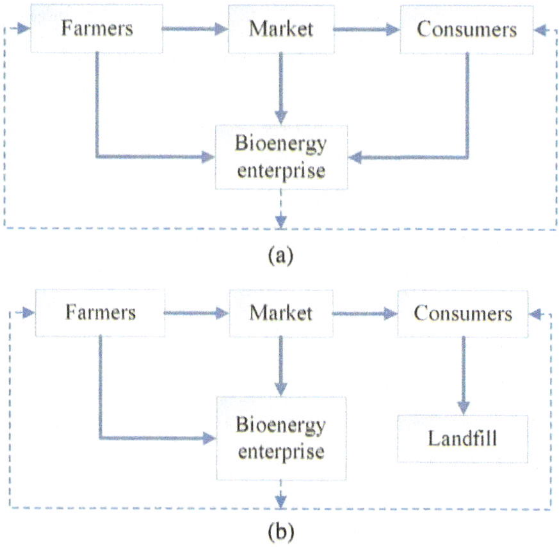

Fig. 2.8 Scenarios of the redesigned supply chain network. **a** Scenario 1 of the redesigned supply chain; **b** Scenario 2 of the redesigned supply chain

market enter the bioenergy enterprise as shown in Fig. 2.8b. Compared to scenario 1, the latter is set based on possible complexities in waste classification and pretreatment due to individual rationalities of consumers, which may lead to less efficiency in agro-waste collection.

The modeling process usually includes four basic steps: structural analysis of the system, drawing the causal loop diagram, drawing the stock and flow diagram, and model validation [2, 103]. To determine the key factors of the system and understand their interrelationships, the following hypotheses of the supply chain design are proposed:

1. The prices of the same product do not vary significantly in different economic regions;
2. The locations of the farms and the markets, as well as the logistical routes between different supply chain nodes are relatively stable;
3. The supply chain only concerns the mobility of only one type of product.

A casual loop diagram is drawn to illustrate the system structure and the interrelationships among key factors. In Fig. 2.9, there are two reinforcing loops and three balancing loops. The first reinforcing loop shows the increase in agro-wastes and the consequent increase in total agro-wastes due to increased market acquisition, which leads to a higher production of biomass energy and a lower consumption of traditional energy; this reduces the supply cost and eventually encourages more market acquisition. As for the second reinforcing loop, the increased production of agro-products leads to more agro-wastes and more total wastes; this stimulates biomass energy production and lowers the farmers' consumption of traditional energy, which in turn reduces the supply cost and expands the production of agro-products.

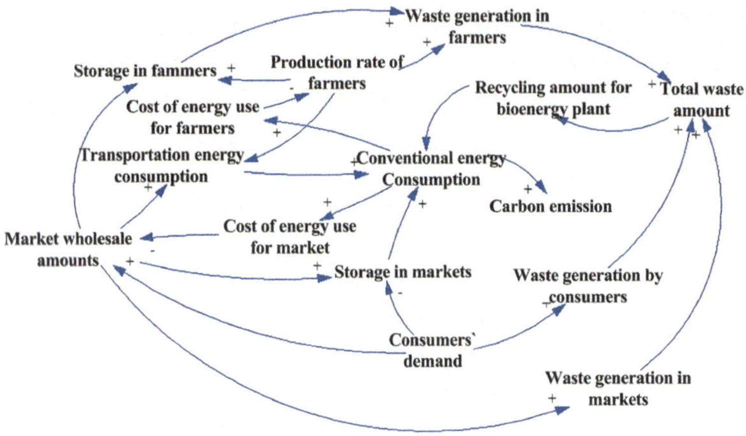

Fig. 2.9 Causal loop diagram of the agro-product supply chain

In terms of the balancing loops, the first is the increased energy consumption caused by increased market acquisitions; this results in increased energy consumption and higher costs of the supply chain, which in turn reduces the amount of market acquisitions. The second loop is the decreased stock of agro-products due to increased market acquisitions, followed by a decreased production of agro-wastes and thus biomass energy; as a result, the use of traditional energy increases, increasing the total cost and eventually decreasing the number of market acquisitions. The last loop is about the increase in the stock of agro-products due to increased market acquisitions; this leads to the increased use of traditional energy and then to higher supply costs, reducing the market acquisitions.

Based on the above analyses, two-feedback-loop clusters are defined as follows:

Reinforcing loop clusters: Marketing acquisition → Marketing waste generation → Total waste generation → Recycling amount of bioenergy plant → Conventional energy consumption → Cost of energy use → Marketing acquisition; Farmers' agro-production → Waste generation in farming stage → Total waste generation → Recycling amount of bioenergy plant → Conventional energy consumption → Cost of energy use → Farmers' agro-production.

Balancing loop clusters: Marketing acquisition → Stock in farmers' agro-production → Waste generation in farming stage → Total waste generation → Recycling amount of bioenergy plant → Conventional energy consumption → Cost of energy use → Marketing acquisition; Marketing acquisition → Energy consumption in Marketing acquisition → Conventional energy consumption → Cost of energy use → Marketing acquisition; Marketing acquisition → Stock of agro-products in marketing stage → Conventional energy consumption → Cost of energy use → Marketing acquisition.

With reference to the causal loop diagram, this study adopts Vensim 6.0 to construct a stock and flow diagram to quantify the interrelationship among the key variables as shown in Fig. 2.10. These variables can be divided into three groups:

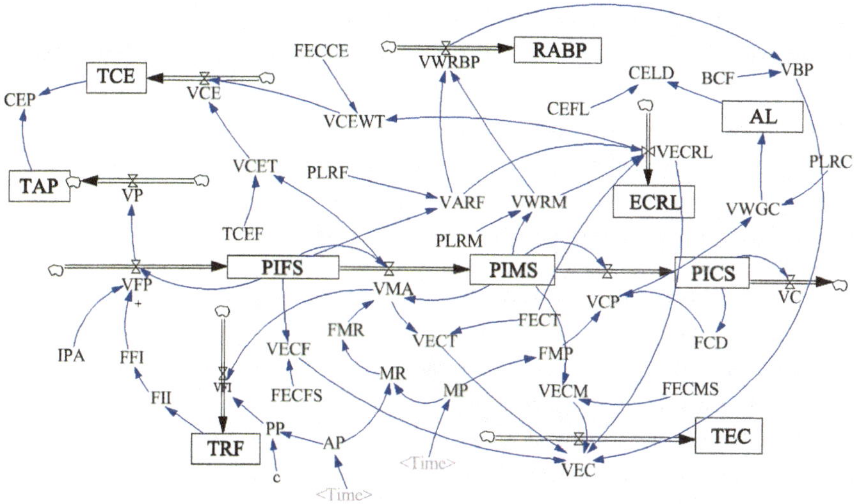

Fig. 2.10 Stock and flow diagram of the agro-product supply chain

state variables, rate variables, and auxiliary variables. State variables are factors that are cumulative over time [103], such as the stock of agro-products. Rate variables are integrals that represent the difference between the input and output rates with respect to time [103]. An example is the integral difference between the stock of agro-products plus the rate of agro-product production and the rate of market acquisition. Auxiliary variables are the intermediates between state and rate variables [103]. An example is the variance of carbon emissions. All variables used in the model and their corresponding categories are listed in Table 2.9. The detailed system dynamics equations are shown in Appendix A.

2.3.2 An Illustrative Case

This section uses yellow chives, which are common in southwestern China, as a typical agro-product for the case study. Due to substantial market demand, an industrial pattern of large-scale chive production has been established. The supply chain consists of a production base, a large wholesale market for agro-products, major consuming cities in the vicinity, a landfill for household waste, and a bioenergy enterprise. Their geographical locations are shown in Fig. 2.11. The production base of the yellow chives is the largest in southwestern China, located in the city of Chengdu in the Sichuan Province, and produces more than 8000 tons annually with a sales volume of approximately 16 million RMB and a damage rate of about 20–30%. The damaged products are transported to a nearby landfill that processes municipal waste. As the concept of green supply chain management is put into practice, the

Table 2.9 Key variables of the system dynamics (SD) model

Key variable	Variable type	Input/output
C (cost)	Constant	Input
FECFS (factor of energy consumption in farming stage)	Constant	Input
FECT (factor of energy consumption in transportation)	Constant	Input
FECMS (factor of energy consumption in marketing stage)	Constant	Input
FECCE (factor of energy consumption converted into carbon emissions)	Constant	Input
CEFL (carbon emissions factor of landfill disposal)	Constant	Input
BCF (biomass-to-energy conversion factor)	Constant	Input
TCEF (factor of carbon emissions in transportation)	Constant	Input
PLRF (product loss rate in farming stage)	Rate	Input
PLRM (product loss rate in marketing stage)	Rate	Input
PLRC (product loss rate of consuming stage)	Rate	Input
IPA (initial production of agro-products)	Constant	Input
FMR (factor of marketing revenue)	Constant	Input
AP (acquisition price)	Constant	Input
MP (marketing price)	Constant	Input
PIFS (product inventory in farming stage)	State	Output
PIMS (product inventory in marketing stage)	State	Output
PICS (product inventory in consuming stage)	State	Output
AL (amount of landfill disposal)	State	Output
TAP (total amount of products)	State	Output
TCE (total carbon emissions)	State	Output
TRF (total revenue of farmers)	State	Output
RABP (recycling amount of bioenergy plant)	State	Output
ECRL (energy consumption of reverse logistics)	State	Output
TEC (total energy consumption)	State	Output
VECT (variation of energy consumption in transportation)	Auxiliary	Output
VECF (variation of energy consumption in farming stage)	Auxiliary	Output
VECM (variation of energy consumption in marketing stage)	Auxiliary	Output
CEP (carbon emissions per product)	Auxiliary	Output
PP (profit per product)	Auxiliary	Output
MR (market revenue)	Auxiliary	Output
FII (Farmers' Income Index)	Auxiliary	Output
FFI (factor of farmers' income)	Auxiliary	Output
FMP (factor for marketing price)	Auxiliary	Output
FCD (factor of consumers' demand)	Auxiliary	Output
VFP (variation of farmers' production)	Auxiliary	Output

(continued)

Table 2.9 (continued)

Key variable	Variable type	Input/output
VMA (variation of marketing acquisition)	Auxiliary	Output
VCP (variation of consumers' purchasing)	Auxiliary	Output
VC (variation consumption)	Auxiliary	Output
VFI (variation of farmers' income)	Auxiliary	Output
VP (variation of total products)	Auxiliary	Output
VCE (variation of carbon emissions)	Auxiliary	Output
VWRBP (variation of waste recycling by bioenergy plant)	Auxiliary	Output
VCEWT (variation of carbon emissions in waste transportation)	Auxiliary	Output
VCET (variation of carbon emissions in transportation)	Auxiliary	Output
VARF (variation of agro-waste recycling from farmers)	Auxiliary	Output
VWRM (variation of waste recycling from market)	Auxiliary	Output
VCEL (variation of carbon emissions in landfill disposal)	Auxiliary	Output
VWGC (variation of waste generation in consuming stage)	Auxiliary	Output
VBP (variation of bioenergy production)	Auxiliary	Output
CELD (carbon emissions of landfill disposal)	Auxiliary	Output
VECRL (variation of energy consumption of reverse logistics)	Auxiliary	Output
VEC (variation of energy consumption)	Auxiliary	Output

agro-waste from the production base is integrated with the surrounding bio-energy enterprises to improve the utilization of agro-waste and reduce energy consumption.

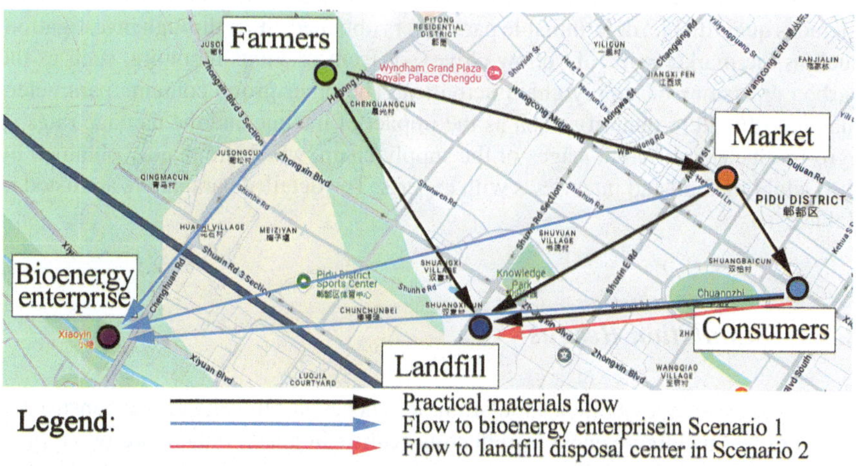

Fig. 2.11 Geographical locations of the case study

Table 2.10 Measurements of the model input

Variable	Value	Source
Cost (C)	0.72 Yuan/kg	Field investigation
FECFS (factor of energy consumption in farming stage)	0.0018 kg diesel/(kg day)	Field investigation
FECMS (factor of energy consumption in marketing stage)	0.0044 kg diesel/(kg day)	Field investigation
FECCE (factor of energy consumption converted into carbon emissions)	3.1388	IPCC [36]
TCEF (factor of carbon emissions in transportation)	0.0347 kg CO_2/kg 100 km	Patel and Kumar [106]
PLRC (products loss rate of consuming stage)	0.3	Delphi method
FECT (factor of energy consumption in transportation)	0.011 kg diesel/kg 100 km	Giordano et al. [107]
CEFL (carbon emissions factor of landfill disposal)	0.2708 kg CO_2/kg waste	Lee et al. [108]
BCF (biomass-to-energy conversion factor)	0.007 kg diesel/kg recycled waste	AI-Hamamre et al. [2]
PLRF (products loss rate in farming stage)	0.25	Delphi method
PLRM (products loss rate in marketing stage)	0.1	Delphi method

The data in this study is mainly collected through field investigation and review of related statistics. Two types of parameters are included in the model. The first group includes quantifiable, determinable parameters obtained either through investigation, such as the market price of yellow chives, or from previous literature, such as the carbon emissions of supply chain activities. The other group contains parameters that are difficult to quantify, such as the impact of income and the damage rates of agro-products at different stages of the supply chain, which are mainly obtained by the Delphi method and interviews with experts. The detailed parameters are listed in Table 2.10.

2.3.3 Simulation Results

Prior to the system dynamics simulation, it is necessary to validate the constructed model to test its effectiveness in reflecting reality. In general, there are two typical methods to ensure the model accuracy, i.e. behavioral sensitivity analysis and historical data comparison, to verify the extent to which the model is consistent with the actual situation [21, 48]. In this section, the latter is used to test the model performance by comparing the discrepancies between the simulated and actual values through error analysis of the model, which in turn quantifies the accuracy of the model. The common practice is to check whether the simulation result of a certain quantitative variable is consistent with the corresponding historical data, e.g. the total income of farmers. The simulation result is compared with the actual statistics of the farmers' income in the recent years. Table 2.11 reflects the differences between the simulation results and the historical statistics for the past seven years, for which the average relative error is shown as 3.93%. The error is considered to be within 15%, which confirms the reliability of the presented model [103].

In the benchmark scenario, the bio-energy enterprises are excluded from the supply chain activities, and all the agro-wastes are delivered to the landfill that processes municipal wastes. Taking 10 years as a simulation cycle, the simulation results of the carbon emissions per unit product in the benchmark scenario are shown in Fig. 2.12.

Figure 2.13 shows the simulation results of the carbon emissions per unit of agro-product under the two proposed scenarios. Compared to the benchmark scenario, both scenarios exhibit a significant decrease in carbon emissions during the simulation

Table 2.11 Comparison of simulated and historical values of farmers' income

Farmers' income	2010	2011	2012	2013	2014	2015	2016
	Ten thousand Yuan						
Simulation result	206.79	220.01	206.19	221.34	216.33	220.48	217.23
Statistical data	205.12	214.32	220.67	212.83	230.59	225.34	229.03
Relative error (%)	0.81	2.65	6.56	4.00	6.18	2.16	5.15

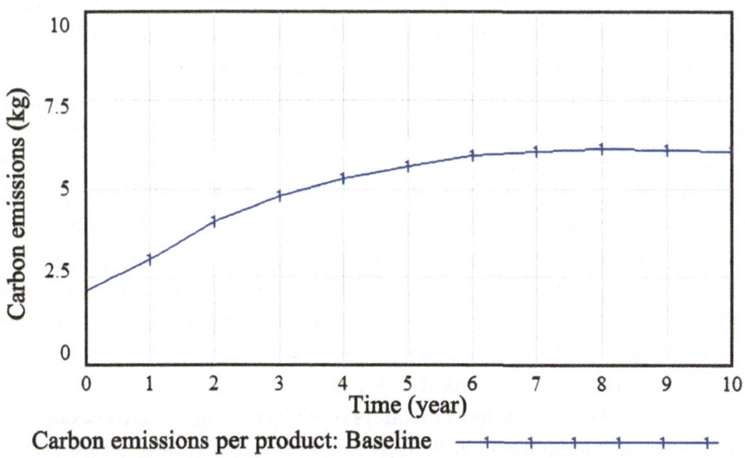

Fig. 2.12 Carbon emissions per unit product in the benchmark scenario

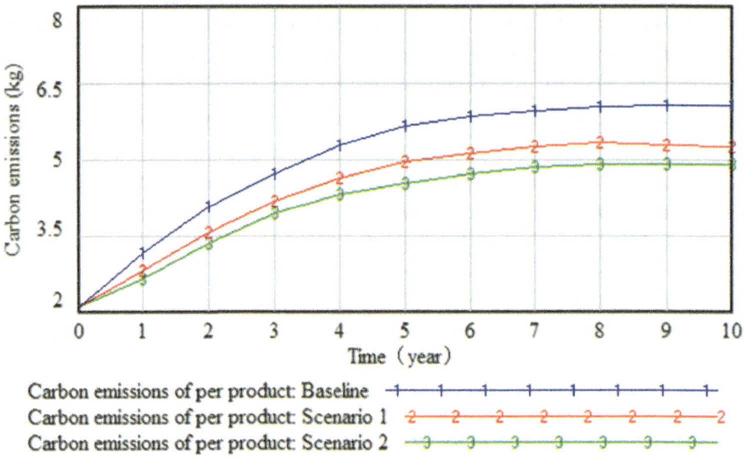

Fig. 2.13 Carbon emissions per unit product under two scenarios

cycle. Scenario 1 shows a cumulative decrease of 14.13%, while scenario 2 shows a cumulative decrease of 18.02%. It is evident that scenario 2 is more effective in reducing carbon emissions. This conclusion indicates that although the recycling and processing of agro-waste generates a greater amount of bioenergy, this approach is undesirable because the dispersion of consumers contributes to an increase in carbon emissions during waste collection and transportation and, consequently, an increase in total emissions.

The above analysis shows that, despite the substantial environmental benefits of a green supply chain, the energy production of the bioenergy enterprises is relatively small due to the external costs of energy savings and emissions reduction. The wide dispersion of the agro-waste leads to high costs of waste collection and logistics, which cripples the management and implementation of the agro-product supply chain. In this context, government policies such as subsidies, green procurement, and tax incentives play an important role in maintaining the stable operation of a green supply chain. Given that scenario 2 has been selected as the preferred mode of operation, this section identifies the optimal incentive mechanism by observing how government subsidies influence the emissions reduction of the supply chain.

Government investments and subsidies are effective in helping bioenergy enterprises improve their operation and development [30]. In this study, it is assumed that after the subsidy, the bioenergy enterprises can choose a high-cost cleaner production plan to increase their energy conversion efficiency to three times the original one. Under such a context, the simulation results indicate an insignificant decrease in carbon emissions per unit product of 6.05%, as shown in Fig. 2.14.

In this scenario, the government considers subsidizing logistics to encourage transportation in the supply chain to shift from fossil energy driven to renewable energy driven. According to Fig. 2.15, the carbon emissions per unit of product in this scenario decrease by 11.29%. Compared to subsidizing the bio-energy enterprises,

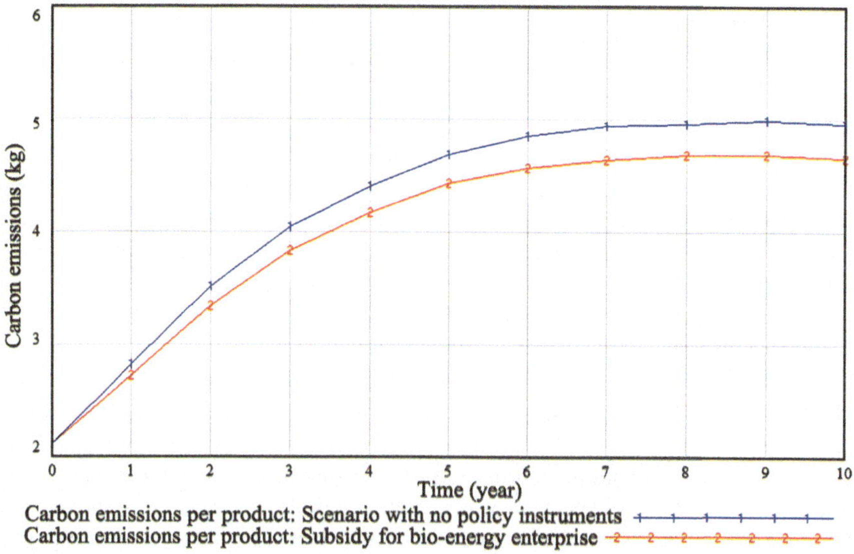

Fig. 2.14 Carbon emissions per unit product under subsidization of bioenergy enterprises

logistics subsidies have significantly more economic and environmental benefits. Therefore, any attempt to optimize the agro-product supply chain can give priority to reducing emissions in the logistics stages.

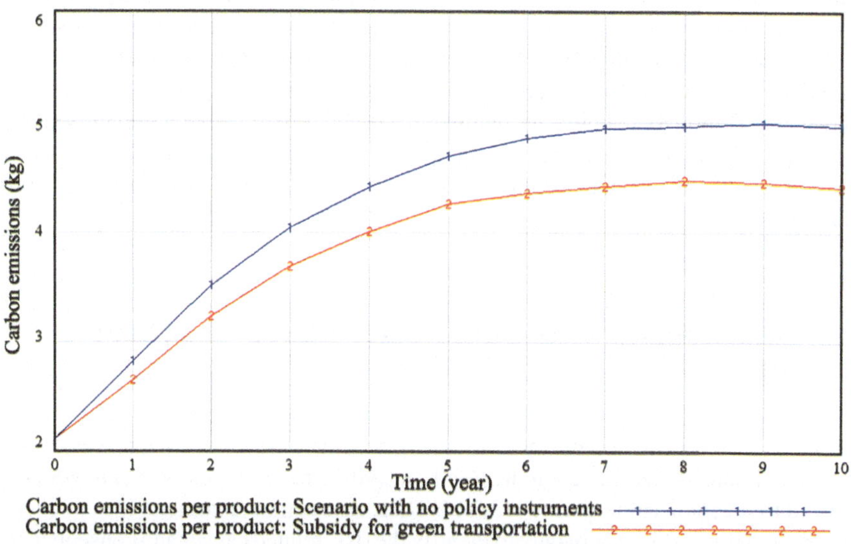

Fig. 2.15 Carbon emissions per unit product under logistic subsidization

2.3.4 Summary

This study redesigns a regional agro-product supply chain by taking bioenergy enterprises into account to reduce ago-wastes and enhance eco-efficiency. Two operational scenarios are created, where scenario 1 assumes that all the agro-waste flows into the bioenergy enterprise, while scenario 2 only considers the agro-wastes produced by farmers and the wholesale market for energy production. System dynamics offers a powerful tool to select the optimal scenario to improve the eco-efficiency of the proposed supply chain. The simulation result shows that scenario 2 has a greater potential in improvement of the eco-efficiency of the agro-product supply chain, particularly in terms of carbon emissions reduction.

The case example is used to demonstrate the application of system dynamics, which is used to select an operations scenario for supply chain management practice to improve its eco-efficiency, using carbon emissions per unit product as an indicator for the simulation. In order to promote such a scenario into management practice, an incentive mechanism was designed based on government subsidy for the bioenergy enterprise or the logistics processes. The simulation result shows that subsidy for the logistics processes is better for facilitating the supply chain operations. For instance, subsidizing the logistic processes can effectively reduce the emissions intensity. This is also an indirect evidence that the carbon emissions of the agro-product supply chain are largely caused by the logistics processes [61]. In comparison, subsidizing bioenergy enterprise is ineffective in terms of emissions reduction. This finding indirectly confirms the research conclusion of Yue et al. [91], which states that the lack of a systematic coordination mechanism within bioenergy enterprises has hindered their power generation capacity and industrial scale, which in turn limits the benefits of adopting high-efficiency energy conversion technologies to reduce carbon emissions. Therefore, it is suggested that sufficient attention should be paid to the coordination and balance among stakeholders of a supply chain, while bioenergy enterprises should be industrialized and expanded to a certain scale to gradually highlight their market competitiveness and eco-efficiency. It is expected that the results can provide insight into the optimization of agro-product supply chain management.

2.4 Vulnerability Assessment for Key Enterprise in Supply Chain

From a disaster formation perspective, safety accidents are the consequences caused by hazard-causing factors combined with hazard-bearing bodies, which are closely related to receptor vulnerability [12, 67]. Thus, assessing the vulnerability of chemical enterprises is critical not only to identify the risk-related causes and consequences of safety accidents, but also to provide insights into risk prevention and control for chemical product supply chain management [46, 68].

Previous vulnerability studies have primarily focused on an assessing the external vulnerability of enterprises, specifically to identify hazard sources and predict the possible consequences of safety accidents [47, 60]. However, the inherent receptivity of hazard-bearing entities to such accidents is often neglected. First, the consequences may differ even when the same hazard-causing factor acts on different hazard-bearing bodies. Second, differences in the the sensitivity and adaptability of hazard-bearing entities to a hazard-causing factor may also lead to substantial differences in their tolerance of safety accidents. For instance, adaptability factors such as employees' safety awareness, the frequency of equipment maintenance, and existing safety measures may reduce the potential damage of a safety accident's possible harm to key hazard-bearing bodies, including the employees, facilities, and the environment [17, 38].

Consequently, this study constructs a composite indicator system for vulnerability assessment based on the disaster-causing factors and hazard-bearing bodies involved in chemical safety accidents. A hybrid model (D-FDM), combining a data envelopment analysis (DEA) and fuzzy decision-making, is constructed to aid in this assessment, where the former primarily measures quantitative indicators, while the latter handles those that are non- or semi-quantitative. These two methods can maximize their individual advantages by reducing information loss and improving the validity of the assessment results [27]. A typical ammonia-producing enterprise located in Southwestern China's Sichuan Province is taken as a case example to verify the assessment model and propose effective measures for risk prevention. It is expected that the study can provide insight into the process safety for the chemical product supply chain.

2.4.1 An Indicator System Design

This study constructs an indicator system for the vulnerability assessment of chemical enterprise based on three dimensions: exposure, sensitivity, and adaptability, as shown in Table 2.12. On the one hand, exposure and sensitivity are classified as physical vulnerability indicators, which include the vulnerability of people, the environment, and equipment [7]. On the other hand, adaptability refers to an enterprise's ability to adapt to security incidents, or more specifically, its inherent resilience during the occurrence of safety incidents [35, 75].

Employee vulnerability is represented by the employee exposure and density [6]. The former refers to the toxic or harmful chemical substances that employees are exposed to and the employees' work hours [74]. The employee exposure is measured based on an exposure-dose assessment, representing the chronic harm to employees due to exposure to toxic substances [83]. Glendon et al. [24] highlighted such an assessment by quantifying the possible terminal health effects of employees when exposed to toxic substances based on the intensity, frequency and duration of exposure, which are given as follows:

Table 2.12 Indicator system of the vulnerability assessment for chemical enterprise

Category	First-Level Index	Second-Level Index
Physical vulnerability	Employees' exposure	Concentration of hazardous materials Work hours
	Employees' density	Number of employees per facility Production facility's size
	Externally environmental hazard	Residential density Number of sensitive areas Number of enterprises nearby
	Internally environmental hazard	Amount of hazardous materials Variety of hazardous materials Latent chemical energy in the hazardous materials
	Equipment identification	Duration of service Operating pressure
	Possible consequence	Number of injuries Number of fatalities Economic loss Environmental loss
Adaptability	Frequency of safety training for employees Frequency of emergency training Close-circuit television monitoring system Enterprise's surveillance Maintenance cycles of equipment	

$$Intake\ (\text{mg/kgd}) = \frac{CA \times IR \times ET \times EF \times ED}{BW \times AT} \tag{2.42}$$

where CA denotes the concentration of air pollutants, or mg/m^3; IR is the inhalation rate, m^3/h; ET refers to the exposure time, h/d; EF is the exposure frequency, d/a; ED signifies the exposure duration, a; BW is the body weight in kg, and AT is the average time of contact, d.

Employee density correlates with the number of employees in production or support facilities, which is measured as follows:

$$\rho = \frac{NP}{S} \tag{2.43}$$

where ρ stands for the density, $capita/\text{m}^2$; NP is the number of employees, $capita$; and S is the size of the facility, m^2.

An enterprise's environmental vulnerability is divided into internal and external environmental hazards. The former (internal environmental hazard) manifests itself as the storage of hazardous materials, which is given as follows [1]:

$$M = \frac{\sum_{i=1}^{n}(Q \times MF)}{n} \tag{2.44}$$

where M is the hazard index; n is the variety of hazardous materials; Q denotes the ratio of the actual storage of hazardous materials to the critical mass; and MF is the hazard coefficient.

The external environmental hazard is measured by the residential density outside the chemical enterprise, including the number of sensitive areas and similar enterprises nearby, which is given as follows [37]:

$$\mathrm{Sv} = \frac{\mathrm{Sr} + \mathrm{St} + \mathrm{Sp}}{3} \tag{2.45}$$

where Sv is the enterprise's externally environmental vulnerability. Further, Sr denotes the residential density, categorized into 4 levels: "high density" (≥ 1500 $capita/\mathrm{km}^2$), "medium density" (600–1500 $capita/\mathrm{km}^2$), "low density" (≤ 600 $capita/\mathrm{km}^2$), and "uninhabited." Each level is assigned a value of 1, 0.75, 0.5, or 0.25, respectively [47]. Moreover, St is the sensitive area index, divided into 4 levels: "high sensitivity" (the number of environmentally sensitive areas is ≥ 5), "medium sensitivity" (≥ 3), "low sensitivity" (≥ 1), and "zero sensitivity" (0), each of which is assigned a value of 1, 0.75, 0.5, or 0.25, respectively [37]. The number of similar chemical enterprises nearby is represented by Sp. When this is ≥ 7, the value is 1, when the number is $\in[5,7]$, $\in[3,5]$, or $\in[1,3]$, the values are assigned as 0.75, 0.5, or 0.25, respectively [37].

Equipment vulnerability is represented by the equipment identification index, measured as follows [35]:

$$\mathrm{H} = \left(1 + \frac{\sum_{\mathrm{i}=1}^{3}\mathrm{Ki}}{3}\right) \times \mathrm{e} \tag{2.46}$$

where H is the equipment identification index;e is the base equipment identification number; Ki is factor influencing equipment identification; and i denotes the sequence number of the influencing factors, with $i = 1, 2, 3$ corresponding to the three influencing factors: the service duration, operating temperature and pressure, and maintenance funds.

In general, the possible consequences are determined by the frequency of accidents and the severity of consequences [10]. However, in actual operations, different influencing factors may have the same risk level, which makes it impossible to distinguish their consequences. Hence, this study introduces the risk level index, represented by the Borda count method, as follows [82]:

$$B_i = \left(N - I_j\right) + \left(N - P_s\right) \tag{2.47}$$

where B_i stands for the Borda ranking of the ith facility; N denotes the number of facilities with a sequence order, and P_s is the sequence order of the consequence

Table 2.13 Classification of adaptability assessment indicators

Indicator	Classification of assessment			
	I	II	III	IV
Frequency of safety training	Quarterly	Semiannually	Annually	None
Frequency of emergency training	Quarterly	Semiannually	Annually	None
Closed-circuit television monitoring system	Within all units	Within production units	Within some production units	None
Enterprise's surveillance	Automated online surveillance	Regular surveillance	Irregular surveillance	None
Equipment maintenance	Compulsory periodic maintenance	Regular maintenance	Irregular maintenance	None

probability. Further, I_j is the sequence order of the consequence intensity, measured as follows:

$$I_j = C_j + (1 + M_j)/2 \tag{2.48}$$

where $C_j = \sum_{r=1}^{j-1} M_r (j > 1)$, $C_1 = 0$; M_j represents the number of facilities with a severity of $Q_j (j \leq 4)$, with Q_1, Q_2, Q_3, and Q_4 denoting severities of I, II, III, and IV, respectively. Moreover,

$$P_s = D_s + (1 + N_s)/2 \tag{2.49}$$

where $D_s = \sum_{r=1}^{s-1} N_r (s > 1)$, $D_1 = 0$;N_s denotes the number of facilities with a probability ranking of $H_s (s \leq 5)$, with H_1, H_2, H_3, H_4, and H_5 representing probabilities of grades A, B, C, D, and E, respectively.

All of the indicators used in this study to describe the four levels of adaptability are qualitative, as shown in Table 2.13.

2.4.2 *Model Construction*

The indicator system for enterprise's vulnerability is composed of the physical vulnerability indicators and the adaptability indicators, where the former is mainly quantitative and the latter is qualitative. Thus, a hybrid D-FDM model is proposed to assess the vulnerability of chemical enterprise in the study. The assessment is divided into two processes, i.e. primary and secondary, as shown in Fig. 2.16. DEA is used to handle with the quantitative indicators, which produces assessment results of physical vulnerability. These results and the data related to the adaptability indicators are then fuzzified simultaneously via membership functions (MFs). The results of fuzzification are taken as the parameters for fuzzy evaluation to obtain the final results, i.e. the comprehensive vulnerability assessment results. The DEA overcomes the subjectivity in fuzzy decision-making, and the fuzzy processing of the DEA results ensures the interrelationships of the assessment results without information loss [49]. The integration of these two methods can lead to the maximization of their advantages and facilitate multi-attributive decision making in vulnerability assessment.

The DEA uses a relative efficiency measurement as the basis for determining the input–output ratio of each decision unit [14]. From the perspective of input–output analysis, the vulnerability of chemical enterprises is indicated by their "disaster efficiency": a higher efficiency means that accidents are more likely to occur and the enterprise is more vulnerable [3, 34]. Suppose there are n facilities in a chemical enterprise as decision-making units (DMU_j),m inputs and s outputs are then randomly assigned to the DMU_j. The input and output vectors are $X_j = \left(X_{1j}, X_{2j} \ldots X_{mj}\right)^T$ and $Y_j = \left(Y_{1j}, Y_{2j} \ldots Y_{sj}\right)^T$, respectively, while the corresponding weight vectors are $v = (v_1, v_2 \ldots v_m)^T$ and $u = (u_1, u_2 \ldots u_s)^T$, respectively,$\theta$ denotes the efficiency index. Thus, the constructed efficiency ratio model is given as follows:

$$\theta = \frac{u^T \cdot Y_j}{v^T \cdot X_j} \tag{2.50}$$

If θ tends toward 1, this indicates that the enterprise has a greater safety vulnerability. The following optimization is generated by taking $\theta \leq 1$ as the constraint condition for all decision units $DMU_j(j = 1, 2, \ldots n)$:

$$\begin{cases} max\frac{u^T Y_j}{v^T X_j} \\ s.t.\frac{u^T Y_j}{v^T X_j} \leq 1, j = 1, 2, \ldots n \\ u_k \geq 0, k = 1, 2, \ldots s \\ v_j \geq 0, j = 1, 2, \ldots m \end{cases} \tag{2.51}$$

According to the Charnes-Cooper transformation,

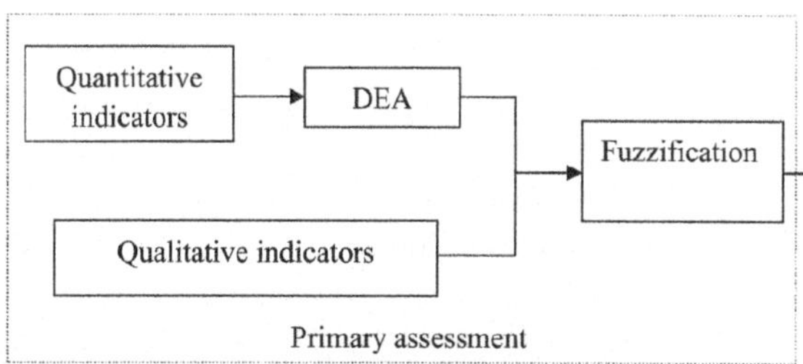

Fig. 2.16 The DEA coupled with fuzzy assessment

$$\begin{cases} t = \frac{1}{v^T X_j} \\ \omega = tv \\ \mu = tu \end{cases} \tag{2.52}$$

and Eq. (2.51) transforms into Eq. (2.53):

$$\begin{cases} maxu^T Y_j = v_{c^2R} \\ s.t.\omega^T X_j - \mu^T Y_j \geq 0, j = 1, 2, \ldots n \\ \omega^T X_j = 1 \\ \omega \geq 0, \mu \geq 0 \end{cases} \tag{2.53}$$

Finally, by substituting the input and output data of n decision units into Eq. (2.50), the efficiency index θ for each assessment unit is obtained.

For the quantitative indicators, fuzzification is performed by using the triangular membership functions to transform the DEA assessment results into corresponding membership degrees, and the procedures are given as follows [78]:

1. Determine the index set U of the assessment object:

$$U = \{u_1, u_2 \ldots u_n\} \tag{2.54}$$

2. Determine the assessment set V:

$$V = \{v_1, v_2 \ldots v_m\} \tag{2.55}$$

3. Solve the membership matrix R, as follows:

$$\mathrm{R} = \begin{bmatrix} \mathrm{r}_{11} & \cdots & \mathrm{r}_{1\mathrm{m}} \\ \vdots & \ddots & \vdots \\ \mathrm{r}_{\mathrm{n}1} & \cdots & \mathrm{r}_{\mathrm{nm}} \end{bmatrix}, 0 \ll \mathrm{rij} \ll 1 \tag{2.56}$$

For the results X_i obtained by DEA, the matrix is developed as follows:

If X_i exceeds the level I standard, or $X_i > V_1$, then the membership degrees of X_i to different levels are as follows:

$$r_{i1} = 1, r_{i2} = r_{i3} = r_{i4} = 0. \tag{2.57}$$

If X_i falls between the level j and j + 1 standards, or $V_j > X_i > V_{j+1}$, then the membership degrees of X_i to different levels are as follows:

$$r_{i(j+1)} = \frac{v_j - x_i}{v_j - v_{j+1}} \tag{2.58}$$

$$r_{i(j)} = 1 - r_{i(j+1)} \tag{2.59}$$

From Eq. (2.58) to (2.59), the typical triangular membership functions are shown in Fig. 2.17.

If X_i is below the level IV standard, or $X_i < V_4$, then the membership degrees of X_i to different levels are as follows:

$$r_{i4} = 1, r_{i1} = r_{i2} = r_{i3} = 0 \tag{2.60}$$

Similarly, the membership functions corresponding to the adaptability assessment indicators (shown in Table 2.13) can be also determined using Eq. (2.57) through Eq. (2.60).

The final results of the comprehensive vulnerability are derived from the fuzzy evaluation. In addition, the objects are ranked by a fuzzy linear transformation based on the principle of maximum membership degree [28, 92]. The procedures are given as follows [26, 92]:

First, the indicators' weighted coefficient vector A should be given as:

$$A = \{a_1, a_2 \ldots a_n\} \tag{2.61}$$

An analytic hierarchy process (AHP) is adopted to assign the weighted values for the indicators, expressed as follows [9, 69]:

1. Construct a judgment matrix and perform paired comparisons of indicators to obtain the relative importance, measured on a scale of one to five;
2. Solve the judgment matrix and use each column's geometric mean to obtain the row vector $w_i = (w_1, w_2 \ldots w_n)^T, i = 1, 2 \ldots, n$; subsequently, normalize w_i to obtain the weighted vector w_i';
3. Compute the consistency index CI and consistency ratio CR from the judgement matrix to test its consistency:

$$CI = \frac{\lambda_{max} - n}{-1} \tag{2.62}$$

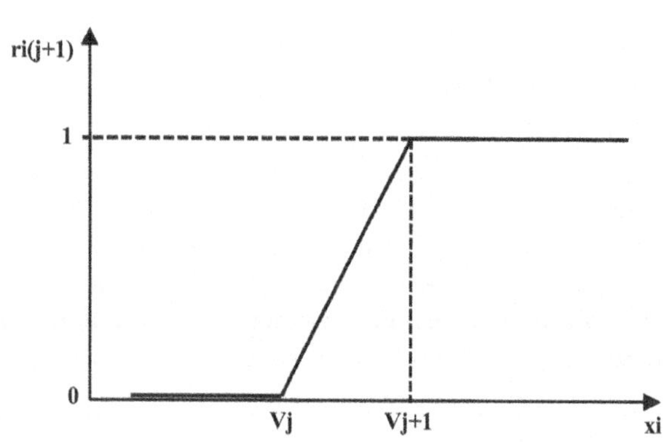

Fig. 2.17 Triangular membership functions used in the study

$$CR = \frac{CI}{RI} \tag{2.63}$$

When $CR \leq 0.1$, the matrix is considered consistent.

Finally, the final assessment vector will be composed: $B = A^{\circ}R$, where R is the membership matrix from Eq. (2.56). The fuzzy operator in this study, "$\circ$," uses ordinary additive operations within bounded constraints:

$$B_k = min\left\{1, \sum_{j=1}^{m} min\left(w_j, r_{jk}\right)\right\}, j = 1, 2, \ldots \tag{2.64}$$

2.4.3 An Illustrative Case

This study focuses on the case of a chemical plant that synthesizes ammonia from natural gas in Sichuan Province to assess its safety vulnerability. Seventeen facilities are located at the industrial site, including major production facilities such as the gas distribution station and the desulfuriation, decarburization and methanation, compression and synthesis, and urea production departments. These facilities also include support departments, such as those related to the desalinated water supply, maintenance, warehousing, and laboratory. Figure 2.18 shows the plant layout.

The input data for the physical vulnerability assessment are primarily obtained through a field survey and a review of related statistics from the case plant. The primary data are processed using Eqs. (2.42) to (2.49), and normalized for input into the MaxDea software for the DEA analysis, as shown in Table 2.14. Regarding its adaptability, five senior management executives specializing in safe production within the case plant are asked to rank the importance of each adaptability indicator. The mean scores are entered into the AHP model to obtain the weightings, as shown in Fig. 2.19.

2.4.4 Results

Figure 2.20 shows the results of the physical vulnerability assessment for the case plant. Clearly, the production facilities are more physically vulnerable than the support facilities. For instance, the desulfurization and decarburization and methanation departments are more vulnerable. This finding parallels the conclusion of Darbra et al.'s [16], that chemical accidents are more likely to occur during the production process.

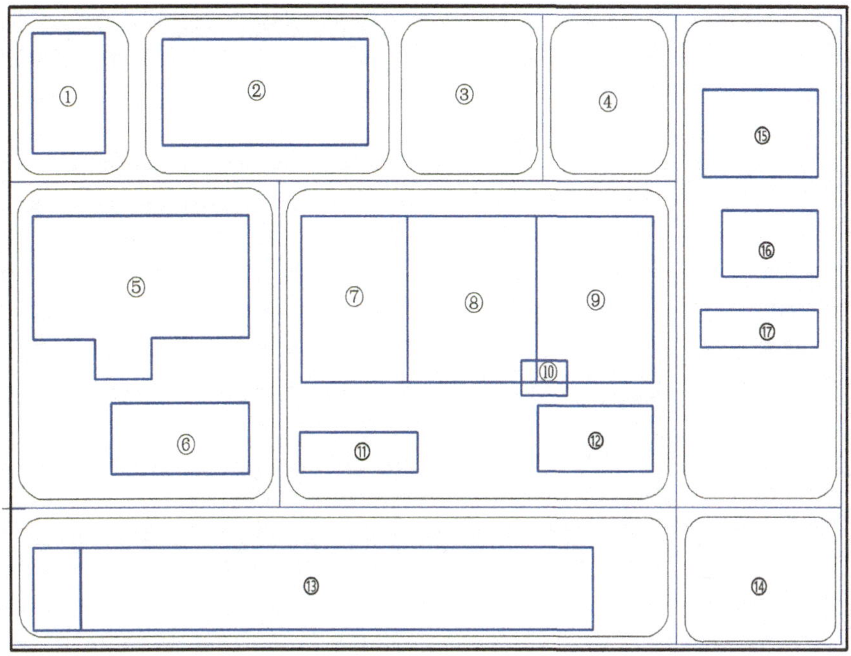

1. Wastewater treatment department 2. Water supply 3. Power supply unit 4. Liquid ammonia storage 5. Urea production 6. Laboratory 7. Compression and synthesis department 8. Decarburization and methanation department 9. Desulfuration department 10. Waste gas treatment 11. Nitrogen generation department 12. Gas distribution station 13. Warehouse 14. Office building 15. Desalted water supply 16. Heat supply 17. Maintenance department

Fig. 2.18 The layout of the investigated chemical plant

The catastrophic efficiency of the heat supply, compression and synthesis, and liquid ammonia storage departments is equal to one, indicating that these three facilities have the greatest safety vulnerabilities. This finding is related to the production nature of the chemical industry. Since the compression and synthesis and liquid ammonia departments are both key facilities along the production line, they contain many equipments related to chemical reactions, and a complex reaction environment [42]. Thus, they may be more vulnerable, and more likely to experience safety accidents. A field survey of the case plant revealed that the heat supply department installed a gas boiler with a capacity of 22 ton/h, which has experienced relatively frequent breakdowns over the past decade due to various failures, including an improper ignition and unstable flame. This shows that the department is physically vulnerable.

Figure 2.21 shows the results of the adaptability assessment for each facility. The heat supply department exhibits extremely high adaptability, accounting for 50% of the total. The compression and synthesis, decarburization and methanation,

Table 2.14 Input data of physical vulnerability indicators

Unit	Employee's exposure	Employee's density	Externally environmental hazard	Internally environmental hazard	Equipment identification	Possible consequence
Gas distribution station	0.189	0.0022	0.417	0	34.5	6
Desulfuration department	0.73	0.0036	0.417	15.57	47.81	8
Decarburization and methanation department	0.19	0.0058	0.417	0.0189	48.21	7
Compression and synthesis department	1.97	0.0025	0.417	6.2	49.58	8
Urea production	2.085	0.0027	0.417	0.056	47.085	8
Desalted water supply	0	0.0021	0.417	0	37.5	7
Nitrogen generation department	0	0.0056	0.417	0	30	6
Maintenance	0	0.002	0.417	0	20	6
Warehouse	0	0.0018	0.417	0	25	6
Laboratory	0	0.0036	0.417	0	20	6
Heat supply	0	0.0017	0.417	0	35	8
Wastewater treatment department	0	0.0021	0.417	0	30	7
Waste gas treatment	0	0.0041	0.417	0	30	7
Liquid ammonia storage	1.877	0.0026	0.417	40	70	9
Water supply	0	0.004	0.417	0	30	6
Power supply unit	0	0.0046	0.417	0	20	6
Office building	0	0.211	0.417	0	20	6

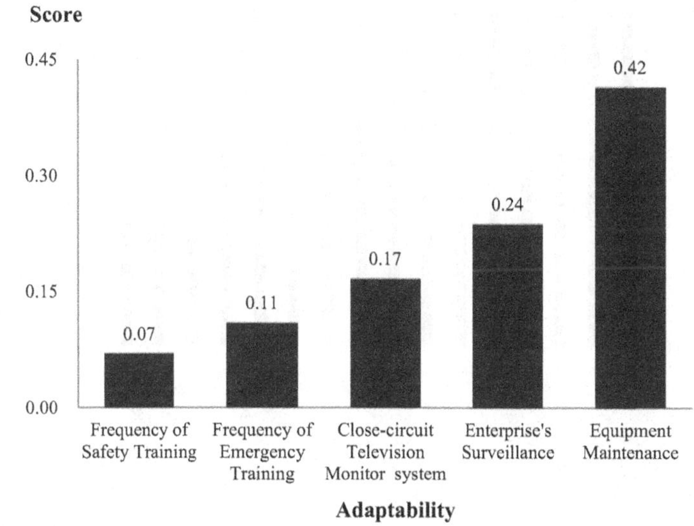

Fig. 2.19 Weight of the indicators for adaptive capacity

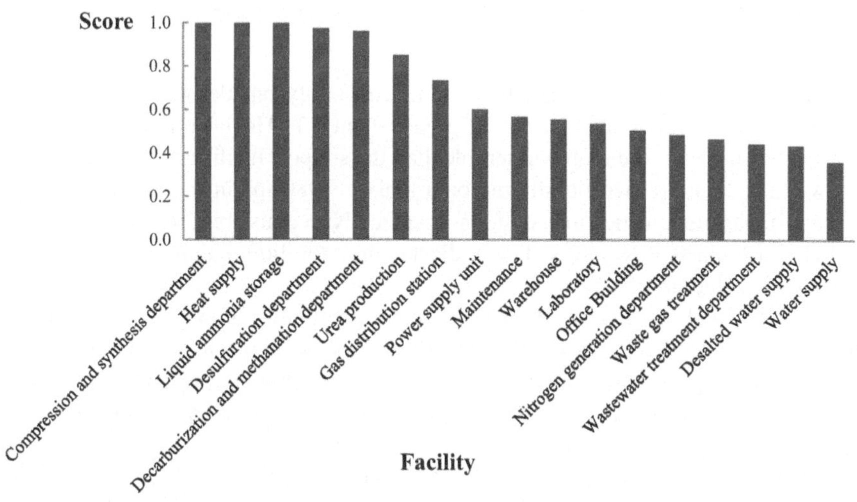

Fig. 2.20 Assessment results of the chemical enterprise's physical vulnerability

liquid ammonia storage, and desulfuration departments indicate high adaptability. As these departments are key facilities in the case plant, more attention is paid to risk prevention, thus reducing the probability of safety accidents.

Figure 2.22 illustrates the results of the comprehensive vulnerability of the case plant. Among all the facilities, the liquid ammonia storage, compression and

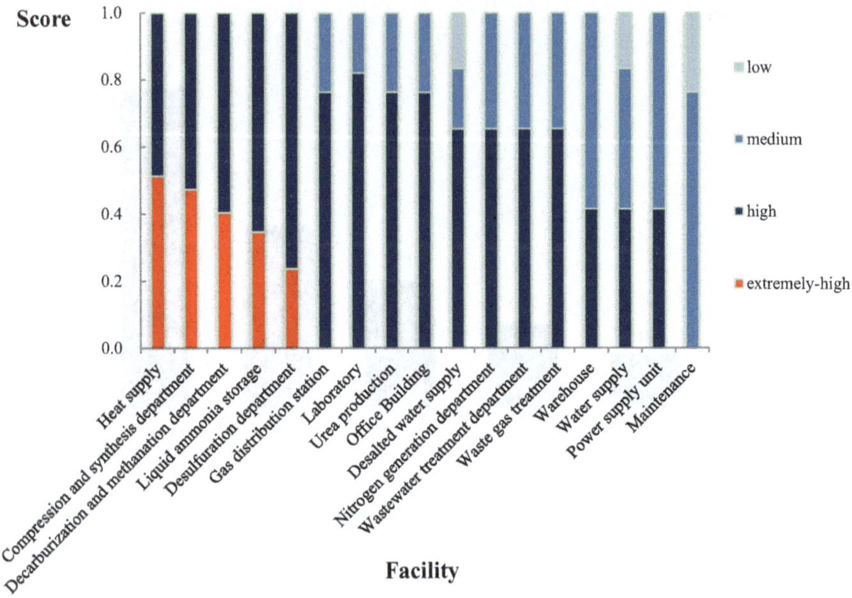

Fig. 2.21 Assessment results of the chemical enterprise's adaptive capacity

synthesis, decarburization and methanation, heat supply, and desulfurization departments all have a disaster efficiency of greater than 0.7. This also indicates that the production facilities are more vulnerable than the support facilities. The field survey shows that there are two liquid ammonia tanks of 980 m^3 in the liquid ammonia storage department, which may emit unorganized NH_3 emissions, resulting in potential health impacts to workers. This indicates that the liquid ammonia department has a high vulnerability.

Figure 2.23 shows that the comprehensive vulnerability of the case plant, which includes the adaptability assessment results, is lower than its physical vulnerability. Despite that, the comprehensive vulnerability of the production facilities remains higher than that of the support facilities. Nevertheless, after incorporating the adaptability assessment results, the scores of the liquid ammonia storage, compression and synthesis, and heat supply departments decrease from 1 to 0.7653, 0.7524, and 0.7484, respectively. This may be closely related to the abundant chemicals used, as well as the complex equipment and environment involved of the production facilities.

From the above results, it can be seen that production facilities are more vulnerable than support facilities . Dahl and Kongsvik [15] indicated that chemical accidents are more likely to occur during chemical production, which indirectly validates the findings of the study. This study further indicated that different amounts of hazardous chemicals were stored in the liquid ammonia storage, desulfurization, and urea production departments of the case plant. Thus, there is a certain risk of leakage exists, which results in a comparatively high vulnerability. Zhang and Zheng [94] verified our results, revealing that 41.5% of the accidents were caused by leakage

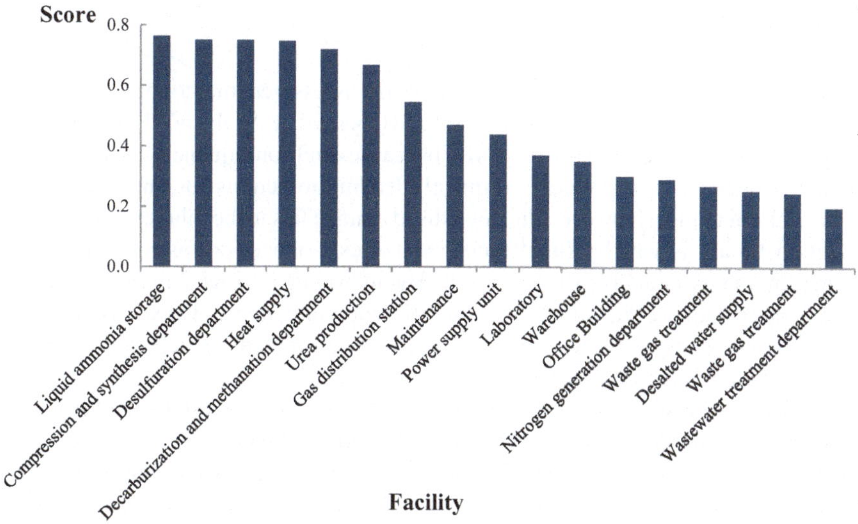

Fig. 2.22 The chemical enterprise's comprehensive vulnerability

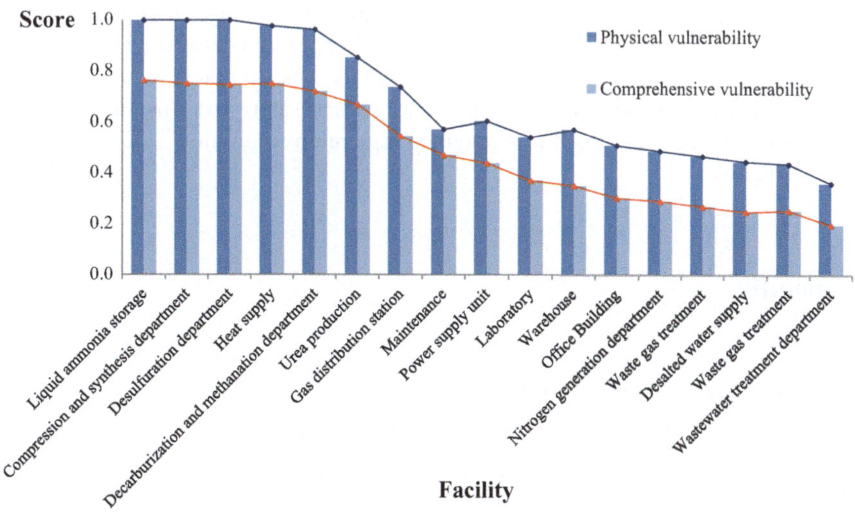

Fig. 2.23 Comparison of the comprehensive and physical vulnerabilities

of these chemicals in fixed facilities. Although the implementation of compensatory measures in the production facilities can reduce safety risks to some extent, the inherent vulnerability of these facilities is still high. Thus, it is necessary to reinforce prevention in order to improve process safety [41].

2.4.5 Summary

This study contributes to the improvement of risk assessment methodology. Taking a typical method of risk assessment "Hazard and Operability Analysis" (HAZOP) as an example, it is effective in identifying possible causes and consequences resulting from deviation of process parameters, ultimately leading to actions for improvement of industrial safety [43]. However, the main disadvantage lies in the inherent structure of qualitative analysis, which is highly dependent on experienced professionals resulting in uneven quality of analysis reports [20]. The hybrid method used in this study fills such a gap by ranking the vulnerability of different production and auxiliary facilities in a chemical enterprise through a quantitative measurement, which can overcome the subjective randomness in risk assessment.

An indicator system for a safety vulnerability assessment is established from the perspective of exposure, sensitivity, and adaptability of enterprise to hazards. The DEA is a hybrid with fuzzy decision-making to develop an assessment model (D-FDM). A chemical plant synthesizing ammonia from natural gas in Sichuan Province, China, is selected as an example to demonstrate the application. Through the identification of facilities that are most likely to fail, this assessment provides insight into improving process safety of the chemical supply chain. The results show that the production facilities, including the liquid ammonia storage, compression and synthesis, decarburization and methanation, heat supply, and desulfurization departments, are more vulnerable than the supporting facilities. Although the introduction of adaptive and compensatory measures can reduce safety risks in all the facilities involved, there is a need to strengthen regulation to promote safe production and operations.

Appendix

VFP = IF THEN ELSE (PIFS < 1000, FFI × IPA, 0)
PIFS(n) = INTEGRAL(VFP(n) − VMA(n) − VARF(n), PIFS(0))
PIFS(0) = 1000
FFI = 1 × FII
FII = VFI/DELAY1I(VFI, 1, 10 million)
VFI (variation of farmers' income) = PP (profit per product) × VMA (variation of marketing acquisition)
PP (profit per product) = AP (acquisition price) – C (cost)
AP (acquisition price) = <time> = ([(0,0) − (10,10)], (1,5.5), (2,6), (3,7.5), (4,8), (5,9), (6,8.8), (7,10), (8,10), (9,10.2), (10,9))
PP (profit per product) = C (cost)

(continued)

(continued)

VMA = IF THEN ELSE (PIFS ≤ 0,0, IF THEN ELSE (PIMS > INVENTORYCAPACITY, 0, 1000 × FMR))
FMR = MP − AP
FMR = FMR/DELAY1I(MR,1, MR(0))
MR(0) = 1
PIMS(n) = INTEGRAL(VMA(N) − VCP(n) − VWRM(n), PIMS(0))
PIMS(0) = 1000
VCP = IF THEN ELSE (PIMS > 0, 1000/(FMP × FCD), 0)
FCD = PICS/DELAY1I(PICS, 1, VCP(0))
VCP(0) = 0
FMP = MP/DELAY1I(MP, 1, MP(0))
MP(0) = 8
MP (marketing price) = <time> = ([(0,0) − (10,10)], (1,8), (2,9.7), (3,10), (4,10.8), (5,11), (6,10.6), (7,11.8), (8,12), (9,13), (10,12.5))
PICS(n) = INTEGRAL(VCFP(N) − VC(n) − VWGC(n), VCFP(0))
VCFP(0) = 100
VECRL = (PLRF × PIFS + PLRM × PIMS) × FECT
ECRL(n) = INTEGRAL(VERAL(n),0)

In scenario 1 of the redesigned supply chain,

AL = 0, CELD = 0

In scenario 2 of the redesigned supply chain,

AL(n) = INTEGRAL(VWGC(n),0) CELD(n) = AL(n) × CEFL
VWGC (variation of waste generation in consuming stage) = VCP (variation of consumers' purchasing) × PLRC (product loss rate of consuming stage)

In scenario 1 of the redesigned supply chain,

VWRBP = VARF + VWRM,
VARF = FIFS × PLRF, VWRM = PLRM × PLMS, VWGC = VCP × PLR

In scenario 2 of the redesigned supply chain,

VWRBP = VARF + VWRM, where VARF = FIFS × PLRF, VWRM = PLRM × PLMS
VEC (variation of energy consumption) = VECF + VECT + VECM + VECRL − VBP
TEC (total energy consumption) = INTEGRAL(VEC(n),VEC(0))
VEC(0) = 0

(continued)

(continued)

VECF = PIFS × FECFS, VECT = FECT × VMA, VECM = PIMS × FECMS
VCE (variation of carbon emissions) = VCET (variation of carbon emissions in transportation) + VCEWT (variation of carbon emissions in waste transportation)
TCE(n) = INTEGRAL(VCE(n),TCE(n − 1))
TCE(0) = 1000
VP (variation of total products) = VFP (variation of farmers' production)
TAP(n) = INTEGRAL(VP(n),0)
CEP (carbon emissions per product) = TCE (total carbon emissions)/TAP (total amount of products)

References

1. AIChE (American Institute of Chemical Engineers) (1994) Dow's fire & explosion index hazard classification guide. American Institute of Chemical Engineers, New York
2. Al-Hamamre Z, Saidan M, Hararah M, Rawajfeh K, Alkhasawneh HE, Al-Shannag M (2017) Wastes and biomass materials as sustainable-renewable energy resources for Jordan. Renew Sustain Energy Rev 67:295–314
3. Alper D, Sinuany-Stern Z, Shinar D (2015) Evaluating the efficiency of local municipalities in providing traffic safety using the data envelopment analysis. Accid Anal Prev 78:39–50
4. Amah E (2012) Corporate culture and organizational effectiveness: a study of the Nigerian banking industry. Eur J Bus Manage 4:212–229
5. Ayoub N, Musharavati F, Pokharel S, Gabbar HA (2015) Risk based life cycle assessment conceptual framework for energy supply systems in large buildings. J Clean Prod 107:291–309
6. Bell ML, Zanobetti A, Dominici F (2013) Evidence on vulnerability and susceptibility to health risks associated with short-term exposure to particulate matter: a systematic review and meta-analysis. Am J Epidemiol 178(6):865–876
7. Birkmann J, Cardona OD, Carreño ML, Barbat AH, Pelling M, Schneiderbauer S, Kienberger S, Keiler M, Alexander D, Zeil P, Welle T (2013) Framing vulnerability, risk and societal responses: the MOVE framework. Nat Hazards 67(2):193–211
8. Bostic RW, Engel KC, McCoy PA, Pennington-Cross A, Wachter SM (2008) State and local anti-predatory lending laws: the effect of legal enforcement mechanisms. J Econ Bus 60(1):47–66
9. Calabrese A, Costa R, Levialdi N, Menichini T (2016) A fuzzy analytic hierarchy process method to support materiality assessment in sustainability reporting. J Clean Prod 121:248–264
10. Casal J (2017) Evaluation of the effects and consequences of major accidents in industrial plants. Elsevier, London
11. Cason TN, Gangadharan L (2014) Promoting cooperation in nonlinear social dilemmas through peer punishment. Exp Econ 18(1):66–88
12. Chakraborty A, Ibrahim A, Cruz AM (2018) A study of accident investigation methodologies applied to the Natech events during the 2011 great east Japan earthquake. J Loss Prev Process Ind 51:208–222
13. Chen Y (2008) The driver of green innovation and green image—green core competence. J Bus Ethics 81:531–543
14. Cook WD, Tone K, Zhu J (2014) Data envelopment analysis: prior to choosing a model. Omega 44:1–4

15. Dahl Ø, Kongsvik T (2018) Safety climate and mindful safety practices in the oil and gas industry. J Safety Res 64:9–36
16. Darbra RM, Palacios A, Casal J (2010) Domino effect in chemical accidents: main features and accident sequences. J Hazard Mater 183(1–3):565–573
17. de Koster RB, Stam D, Balk BM (2011) Accidents happen: the influence of safety-specific transformational leadership, safety consciousness, and hazard reducing systems on warehouse accidents. J Oper Manag 29(7–8):753–765
18. Deutz P, Neighbour G, McGuire M (2010) Integrating sustainable waste management into product design: sustainability as a functional requirement. Sustain Dev 18:229–239
19. Dubey R, Gunasekaran A, Ali SS (2015) Exploring the relationship between leadership, operational practices, institutional pressures and environmental performance: a framework for green supply chain. Int J Prod Econ 160:120–132
20. Dunjó J, Fthenakis V, Vílchez JA, Arnaldos J (2010) Hazard and operability (HAZOP) analysis. A literature review. J Haz Mater 173(1–3):19–32
21. Elsawah S, Pierce SA, Hamilton SH, Van Delden H, Haase D, Elmahdi A, Jakeman AJ (2017) An overview of the system dynamics process for integrated modelling of socio-ecological systems: lessons on good modelling practice from five case studies. Environ Model Softw 93:127–145
22. Eltayeb TK, Zailani S, Ramayah T (2011) Green supply chain initiatives among certified companies in Malaysia and environmental sustainability: investigating the outcomes. Resour Conserv Recycl 55:495–506
23. Filieri R (2015) From market-driving to market-driven: an analysis of Benetton's strategy change and its implications for long-term performance. Mark Intell Plan 33:238–257
24. Glendon AI, Clarke S, McKenna E (2016) Human safety and risk management, 2nd edn. CRC Press, Boca Raton
25. Gualandris J, Klassen RD, Vachon S, Kalchschmidt M (2015) Sustainable evaluation and verification in supply chains: aligning and leveraging accountability to stakeholders. J Oper Manag 38:1–13
26. Guo L, Gao J, Yang J, Kang J (2009) Criticality evaluation of petrochemical equipment based on fuzzy comprehensive evaluation and a BP neural network. J Loss Prev Process Ind 22(4):469–476
27. Han Y, Geng Z, Zhu Q, Qu Y (2015) Energy efficiency analysis method based on fuzzy DEA cross-model for ethylene production systems in chemical industry. Energy 83:685–695
28. Han L, Song Y, Duan L, Yuan P (2015) Risk assessment methodology for Shenyang Chemical Industrial Park based on fuzzy comprehensive evaluation. Environ Earth Sci 73(9):5185–5192
29. Hazen BT, Boone CA, Ezell JD, Jones-Farmer LA (2014) Data quality for data science, predictive analytics, and big data in supply chain management: an introduction to the problem and suggestions for research and applications. Int J Prod Econ 154:72–80
30. He Y, Xu Y, Pang Y, Tian H, Wu R (2016) A regulatory policy to promote renewable energy consumption in China: review and future evolutionary path. Renew Energy 89:695–705
31. Herbert GJ, Krishnan AU (2016) Quantifying environmental performance of biomass energy. Renew Sustain Energy Rev 59:292–308
32. HSE (Health & Safety Executive) (1988) The tolerability of risk from nuclear powerstations. http://www.hse.gov.uk/nuclear/tolerability.pdf
33. Hu (2009) Environmental risk assessment and case study. China Environmental Science Press, Beijing (in Chinese)
34. Huang J, Liu Y, Ma L, Su F (2013) Methodology for the assessment and classification of regional vulnerability to natural hazards in China: the application of a DEA model. Nat Hazards 65(1):115–134
35. Huang L, Wan W, Li F, Li B, Yang J, Bi J (2011) A two-scale system to identify environmental risk of chemical industry clusters. J Hazard Mater 186(1):247–255
36. IPCC (2006) Climate change 2006: synthesis report: contribution of working groups I, II and III to the fourth assessment report. http://www.ipccnggip.iges.or.jp/public/2006gl/pdf/5_Volume5/V5_5_Ch5_IOB.pdf

37. Jia Q, Huang L, Yuan ZW, Zhang XF (2010) Assessment and management of accidental environmental risks in the petrochemical industry. Acta Sci Circum 30(7):1510–1517 (in Chinese with English abstract)
38. John A, Yang Z, Riahi R, Wang J (2016) A risk assessment approach to improve the resilience of a seaport system using Bayesian networks. Ocean Eng 111:136–147
39. Jonkman J, Kanellopoulos A, Bloemhof JM (2019) Designing an eco-efficient biomass-based supply chain using a multi-actor optimisation model. J Clean Prod 210:1065–1075
40. Kane G (2010) The three secrets of green business: Unlocking competitive advantage in a low carbon economy. Earthscan, London
41. Khan F, Hashemi SJ, Paltrinieri N, Amyotte P, Cozzani V, Reniers G (2016) Dynamic risk management: a contemporary approach to process safety management. Curr Opin Chem Eng 14:9–17
42. Kidam K, Hurme M (2013) Analysis of equipment failures as contributors to chemical process accidents. Process Safety Environ Protect Trans Chem Eng Part B 91(1):61–78
43. Kościelny JM, Syfert M, Fajdek B, Kozak A (2017) The application of a graph of a process in HAZOP analysis in accident prevention system. J Loss Prev Process Ind 50:55–66
44. Lee K, Kwon HM, Cho S, Kim J, Moon I (2016) Improvements of safety management system in Korean chemical industry after a large chemical accident. J Loss Prev Process Ind 42:6–13
45. Lee KH (2011) Integrating carbon footprint into supply chain management: the case of Hyundai Motor Company (HMC) in the automobile industry. J Clean Prod 19:1216–1223
46. Leith HM, Piper JW (2013) Identification and application of security measures for petrochemical industrial control systems. J Loss Prev Process Ind 26(6):982–993
47. Li F, Bi J, Huang L, Qu C, Yang J, Bu Q (2010) Mapping human vulnerability to chemical accidents in the vicinity of chemical industry parks. J Hazard Mater 179(1–3):500–506
48. Li Z, Shen GQ, Alshawi M (2014) Measuring the impact of prefabrication on construction waste reduction: an empirical study in China. Resour Conserv Recycl 91:27–39
49. Liu HC, Liu L, Liu N (2013) Risk evaluation approaches in failure mode and effects analysis: a literature review. Expert Syst Appl 40(2):828–838
50. Liu J, Ma Z, Zhang Y, Song L, Li W, Li Y (2014) Key methane emission factors from municipal solid waste landfill treatment in China. Res Environ Sci 27(9):975–980 (in Chinese with English abstract)
51. Liu L, Ye J, Zhao Y, Zhao E (2015) The plight of the biomass power generation industry in China–a supply chain risk perspective. Renewable Sustainable Energy Reviews 49:680–692
52. Longoni A, Cagliano R (2015) Environmental and social sustainability priorities: their integration in operations strategies. Int J Oper Prod Manag 35:216–245
53. Mangla SK, Kumar P, Barua MK (2014) A flexible decision framework for building risk mitigation strategies in green supply chain using SAP–LAP and IRP approaches. Glob J Flex Syst Manag 15:203–218
54. Mezher T, Ajam M (2006) Integrating quality, environmental and supply chain management systems into the learning organisation. In: Sarkis J (ed) Greening the supply chain. Springer, London, pp 67–85
55. Montewka J, Ehlers S, Goerlandt F, Hinz T, Tabri K, Kujala P (2014) A framework for risk assessment for maritime transportation systems—a case study for open sea collisions involving RoPax vessels. Reliab Eng Syst Saf 124:142–157
56. Montgomery DC (2013) Introduction to statistical quality control. Wiley, New York
57. Montmartin B, Herrera M (2015) Internal and external effects of R&D subsidies and fiscal incentives: empirical evidence using spatial dynamic panel models. Res Policy 44(5):1065–1079
58. Motghare KA, Rathod AP, Wasewar KL, Labhsetwar NK (2016) Comparative study of different waste biomass for energy application. Waste Manage 47:40–45
59. Mukhopadhyay A, Goswami A (2014) Economic production quantity models for imperfect items with pollution costs. Syst Sci Control Eng 2(1):368–378
60. Necci A, Cozzani V, Spadoni G, Khan F (2015) Assessment of domino effect: state of the art and research needs. Reliab Eng Syst Saf 143:3–18

61. Nguyen L, Cafferty KG, Searcy EM, Spatari S (2014) Uncertainties in life cycle greenhouse gas emissions from advanced biomass feedstock logistics supply chains in Kansas. Energies 7:7125–7146
62. Okabe M, Ohtani H (2009) Risk estimation for industrial safety in raw materials manufacturing. J Loss Prev Process Indus 22:176–181
63. Park HS, Behera SK (2014) Methodological aspects of applying eco-efficiency indicators to industrial symbiosis networks. J Clean Prod 64:478–485
64. Patrizio P, Leduc S, Chinese D, Kraxner F (2017) Internalizing the external costs of biogas supply chains in the Italian energy sector. Energy 125:85–96
65. Porter ME, Kramer MR (2006) The link between competitive advantage and corporate social responsibility. Harv Bus Rev 84:78–92
66. Rathnayaka S, Khan F, Amyotte P (2014) Risk-based process plant design considering inherent safety. Saf Sci 70:438–464
67. Reniers GL, Cremer K, Buytaert J (2011) Continuously and simultaneously optimizing an organization's safety and security culture and climate: the Improvement diamond for excellence achievement and leadership in safety & security (IDEAL S&S) model. J Clean Prod 19(11):1239–1249
68. Reniers G, Amyotte P (2012) Prevention in the chemical and process industries: future directions. J Loss Prev Process Ind 25(1):227–231
69. Saaty T (1980) The analytic hierarchy process: planning, priority setting, resources allocation. McGraw, New York
70. Santos FM (2012) A positive theory of social entrepreneurship. J Bus Ethics 111(3):335–351
71. Scenna NJ, Santa Cruz AS (2005) Road risk analysis due to the transportation of chlorine in Rosario city. Reliab Eng Syst Saf 90:83–90
72. Scott L, Vigar-Ellis D (2014) Consumer understanding, perceptions and behaviours with regard to environmentally friendly packaging in a developing nation. Int J Consum Stud 38(6):642–649
73. Shaharudin MR, Govindan K, Zailani S, Tan KC, Iranmanesh M (2017) Product return management: linking product returns, closed-loop supply chain activities and the effectiveness of the reverse supply chains. J Clean Prod 149:1144–1156
74. Shi W, Zeng W (2013) Genetic k-means clustering approach for mapping human vulnerability to chemical hazards in the industrialized city: a case study of Shanghai, China. Int J Environ Res Public Health 10(6):2578–2595
75. Shirali GHA, Motamedzade M, Mohammadfam I, Ebrahimipour V, Moghimbeigi A (2012) Challenges in building resilience engineering (RE) and adaptive capacity: a field study in a chemical plant. Process Safety Environ Protect Trans Inst Chem Eng Part B 90(2):83–90
76. Sinclair P, Cohen B, Hansen Y, Basson L, Clift R (2015) Stakeholder engagement within the sustainability assessment of bioenergy: case studies in heat, power and perennial and annual crops from the UK. Biomass Bioenerg 73:11–22
77. Smith P (2007) The relationship between management accounting, profitability and operations in an uncertain world: evidence from literature and practice. PhD thesis, Cranfield University, UK
78. Song Z, Zhu H, Jia G, He C (2014) Comprehensive evaluation on self-ignition risks of coal stockpiles using fuzzy AHP approaches. J Loss Prev Process Ind 32:78–94
79. Stewart WH Jr, Carland JC, Carland JW, Watson WW, Sweo R (2003) Entrepreneurial dispositions and goal orientations: a comparative exploration of United States and Russian entrepreneurs. J Small Bus Manag 41:27
80. The Central People's Government of the People's Republic of China (2007) Regulation of production safety accident report and investigation of China. http://www.gov.cn/zwgk/2007-04/19/content_588577.htm
81. Tonelli F, Evans S, Taticchi P (2013) Industrial sustainability: challenges, perspectives, actions. Int J Bus Innov Res 7:143–163
82. Tsai CF, Hu YH, George Ke SW (2014) A Borda count approach to combine subjective and objective based MIS journal rankings. Online Inf Rev 38(4):469–483

83. Van Leeuwen CJ, Vermeire TG (2007) Risk assessment of chemicals: an introduction, 2nd edn. Springer, Dordrecht
84. Veleva V, Todorova S, Lowitt P, Angus N, Neely D (2015) Understanding and addressing business needs and sustainability challenges: lessons from Devens eco-industrial park. J Clean Prod 87:375–384
85. Wang HF, Gupta SM (2011) Green supply chain management: product life cycle approach. McGraw-Hill, New York
86. Wang JJ, Jing YY, Zhang CF, Zhao JH (2009) Review on multi-criteria decision analysis aid in sustainable energy decision-making. Renew Sustain Energy Rev 13(9):2263–2278
87. Wang J, Dai J (2017) Sustainable supply chain management practices and performance. Ind Manag Data Syst 118(1):2–21
88. Wang S, Fan J, Zhao D, Wu Y (2015) The impact of government subsidies or penalties for new-energy vehicles: a static and evolutionary game model analysis. JTEP 49(1):98–114
89. Xue C, Wang X (2017) Study on government subsidy decision-making of straw power generation supply chain. Proc Eng 174:211–218
90. Yang HJ (2013) Low carbon supply chain management. Science Press, Beijing (in Chinese)
91. Yue D, You F, Snyder SW (2014) Biomass-to-bioenergy and biofuel supply chain optimization: overview, key issues and challenges. Comput Chem Eng 66:36–56
92. Zadeh LA (1965) Fuzzy sets. Inf Control 8(3):338–353
93. Zhang G, Liu P, Gao X, Liu M (2014) Companies' behavior of carbon emission reduction at the risk of oil price volatility. Proc Comput Sci 31:291–298
94. Zhang HD, Zheng XP (2012) Characteristics of hazardous chemical accidents in China: a statistical investigation. J Loss Prev Process Ind 25(4):686–693
95. Zhao R, Neighbour G, Deutz P, McGuire M (2012) Materials selection for cleaner production: an environmental evaluation approach. Mater Des 37:429–434
96. Zhao R, Neighbour G, Han J, McGuire M, Deutz P (2012) Using game theory to describe strategy selection for environmental risk and carbon emissions reduction in the green supply chain. J Loss Prev Process Ind 25:927–936
97. Zhao R, Huang T, McGuire M (2012) From a literature review to an alternative treatment system for landfill gas and leachate. Challenges 3:278–289
98. Zhao Y, Trejo-Pech CO, Weldon RN (2012) The profitability of the US food supply chain: financial indicators, cross-section and time-series effects. J Appl Bus Res 29:23–34
99. Zhao R, Neighbour G, McGuire M, Deutz P (2013) A software based simulation for cleaner production: a game between manufacturers and government. J Loss Prev Process Ind 26:59–67
100. Zhao R, Peng D, Li Y (2015) An interaction between government and manufacturer in implementation of cleaner production: a multi-stage game theoretical analysis. Int J Environ Res 9(3):1069–1078
101. Zhao R, Zhou X, Han J, Liu C (2016) For the sustainable performance of the carbon reduction labeling policies under an evolutionary game simulation. Technol Forecast Soc Chang 112:262–274
102. Zhao R, Xi B, Liu Y, Su J, Liu S (2017) Economic potential of leachate evaporation by using landfill gas: a system dynamics approach. Resour Conserv Recycl 124:74–84
103. Zhao R, Zhou X, Jin Q, Wang Y, Liu C (2017) Enterprises' compliance with government carbon reduction labelling policy using a system dynamics approach. J Clean Prod 163:303–319
104. Zhu QH, Dou YJ (2007) Evolutionary game model between governments and core enterprises in greening supply chains. Syst Eng Theory Pract 27:85–89
105. Zhu QH, Sarkis J (2006) An inter-sectoral comparison of green supply chain management in China: drivers and practices. J Clean Prod 14:472–486

106. Patel HK, Kumar S (2017) Experimental analysis on performance of diesel engine using mixture of diesel and bio-diesel as a working fuel with aluminum oxide nanoparticle additive. Therm Sci Eng Prog 4: 252–258
107. Giordano A, Fischbeck P, Matthews HS (2018) Environmental and economic comparison of diesel and battery electric delivery vans to inform city logistics fleet replacement strategies. Transp Res D Transp Environ 64: 216-229
108. Lee U, Han J, Wang M (2017) Evaluation of landfill gas emissions from municipal solid waste landfills for the life-cycle analysis of waste-to-energy pathways. J Clean Prod 166: 335–342.

Chapter 3
Interaction of Stakeholders Involved in Sustainable Consumption and Production

3.1 Introduction

From a supply chain perspective, enterprises and consumers play a key role in sustainable consumption and production. In this case, they may have bounded rationality, whose decision making is limited by information, cognition, and time [48]. For instance, consumers may lose confidence in new products due to their purchase experiences, which leads to some degree of risk perception [61]. Second, rational individuals aims to maximize their self-interest [30]. This further suggests that enterprises and consumers do not have an enforceable commitment, as the former prefers to make business decisions based on controlled resources rather than cooperating with the latter [11]. While in theory, consumers drive the open market, i.e. they incentivize the sale of attractive and competitive products that meet a perceived need, governments, for good reason, often restrict the market through legislation or regulation, sometimes with unintended consequences. This Chapter examines one such consequence that is particularly relevant to environmental management in the context of sustainable consumption and production, i.e. how governments can provide a reasonable incentive and restraint mechanism to encourage manufacturers to improve their environmental performance.

Both government and manufacturers can be regarded as the key 'actors' involved in the sustainable consumption and production. There is potentially a large area of common interest between government and manufacturers, provided that regulations are designed to achieve their desired outcomes without undesirable unintended consequences [62]. Government is leading society towards sustainability, including contributions to environmental protection, economic development and public satisfaction [31, 57]. With regard to the cleaner production, the government aims at encouraging demand for environmentally friendly products and services, helping business reduce lifecycle environmental impact, and driving innovation, research and design for product sustainability [50, 54]. These actions are mainly enforced by well-designed environmental policies, investments in technological innovation,

R. Zhao, *Multi-criteria Decision-Making Approaches to Sustainable Consumption and Production*, Urban Sustainability, https://doi.org/10.1007/978-981-96-2775-2_3

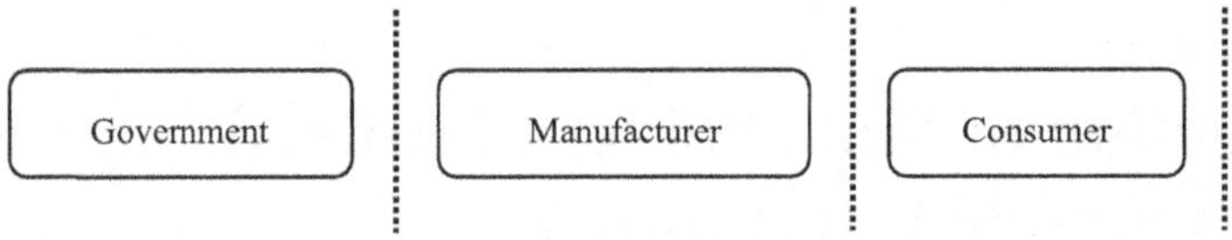

Fig. 3.1 Boundaries of possible game scenarios

allowances for equipment upgrades etc. [12, 14]. While corporations generally respond to a range of signals from consumers, governments within their supply chain, international treaties, regulations etc. [59, 72], the Chapter provides an initial insight into how they might respond to game scenarios with and without government regulations. Game theory can be used to test a model that takes such actions into account and provides an approach to determining the possible dominant strategy in the market.

There are a number of stakeholders involved in a supply chain network, including suppliers, manufacturers, retailers, and consumers, who may face conflicting objectives when making decisions [27]. Game theory is a useful tool to address this dilemma and strengthen coordination among the stakeholders. The game situation can be defined by three key boundaries in the supply chain, government | manufacturer, and manufacturer | consumer, government | consumer, as show in Fig. 3.1. It seems logical to apply game theory first to the dynamic boundary of government | manufacturer, which establishes the necessary regulation for industrial production at an early stage. While an observer might conclude that there is only one game, it is suggested that there are, in reality, several related, but largely independent games. Thus, Sect. 3.2 shows how game theory offers an approach to modeling the behaviour of manufacturer under different policy scenarios in implementation of cleaner production. On this basis, Sect. 3.3 focuses on simulating the possible actions of government and manufacturers in the context of a 'game', in order to achieve more environmentally friendly products. Section 3.4 presents the game theory approach to strategy selection for reducing environmental risk and carbon footprint by manufacturers, first without government supervision, and then with government supervision. Section 3.5 acknowledges the significance of the public's role in consumption to promote sustainable development, and presents an approach by a hybrid of data mining with multiple attribute decision making to identify the optimal mode for last-mile travel choice.

3.2 A Multi-stage Game Theoretical Analysis

A generic game model based on a 'Two-person Non-cooperative' static game is created to allow different strategic actions to be tested by stages, and to aid decision making by selecting the optimal strategy for both the manufacturer and the policy maker to reduce lifecycle based environmental impacts while maximizing the economic benefits. In addition, a case example is given to help a tombarthite manufacturer select the optional clean technology to improve the existing process of production, which provides a useful insight into the application of game theory.

Most of the previously mentioned game theoretic analysis are based on mathematical optimizations to help manufacturers improve their productivity, which are quite useful in informing our approach. However, few of them have addressed the issues of interaction between government and manufacturer raised in the cleaner production. Especially, the results derived from the game models with complex equations containing large parameters or data, may not be easily understood by the stakeholders involved in a game, thus limiting their wider application [60]. This study aims at providing a simple game scenario to model the possible actions of local government and manufacturers in implementation of cleaner production. The game parameters are considered to be generic to help the stakeholders better understand the interaction.

3.2.1 Model Construction

The game has two players: the government (G) and the manufacturer (M). Each of them has two available strategies. For the government, the strategies are: {I, NI} which indicate whether the government should inspect (I) or not inspect (NI) the manufacturer by choosing clean technology to improve the manufacturing process. Accordingly, two strategies for the manufacturer are {IMP, NIMP}. 'IMP' indicates that the manufacturer decides to improve product quality and reduce environmental impact by implementing cleaner production, whilst 'NIMP' means not to implement. Table 3.1 shows the payoff matrix for the two players with different strategic actions.

In Table 3.1, 'R' denotes the possible revenue expected by the government from the implementation of clean technology, 'S' denotes the government subsidy, e.g. equipment subsidy and tax break, to help the manufacturer upgrade the existing production by selection of clean technology, 'C' the cost to the manufacturer to comply with the standard of clean technology, 'E' the expected economic benefit to the manufacturer in using the clean technology, 'P' the economic penalty for the manufacturer's environmentally unfriendly behavior, 'L' the possible loss (L) in terms of the lifecycle based environmental impact of the manufacturer, e.g. waste emission, wastewater discharge, carbon emissions, etc.

Generally speaking, a game is categorized as cooperative or non-cooperative [52]. Their differences can be distinguished by whether agreements or commitments in

Table 3.1 Payoff matrix for government (G) and manufacturers (M)

G	M	
	IMP	NIMP
I	$R-S, S-C+E$	$P-L, -P$
NI	$R, -C+E$	$-L, 0$

a game are enforceable [20]. In this game, both the government and the manufacturer are assumed to be rational, seeking to maximize their self-interest while acting simultaneously, i.e. to maximize their payoffs. Thus, the game can be considered static and non-cooperative [20]. The solution of a game is determined by the "Nash Equilibrium", which is the unique prediction with a best response from the possible strategic actions that each player can choose [18, 21].

3.2.2 A Multi-stage Game Modeling

Stage 1

If the payoff that the manufacturer chooses to implement cleaner production (IMP) is less than that of Non-implementation (NIMP), the following equalities apply:

$$E - C + S < -P \tag{3.1}$$

$$E - C < 0 \tag{3.2}$$

The above two equations can be transformed as $E + S < C - P$, reflecting that the economic return (E) and government subsidy (S) cannot cover the cost ©) Combined with $P - L >- L$, the dominant strategy is that the manufacturer is not willing to implement cleaner production (NIMP), while the government should inspect the manufacturing process (I). Here, a dominant strategy can be expressed as a unique pair of equilibrium (NIMP-I) for this game, as shown in Fig. 3.2. In addition, this scenario indicates that manufacturer may prefer to pay the penalties rather than upgrade the existing production due regard to the cost consideration and insufficient supervision from the government, e.g. a lower economic sanction.

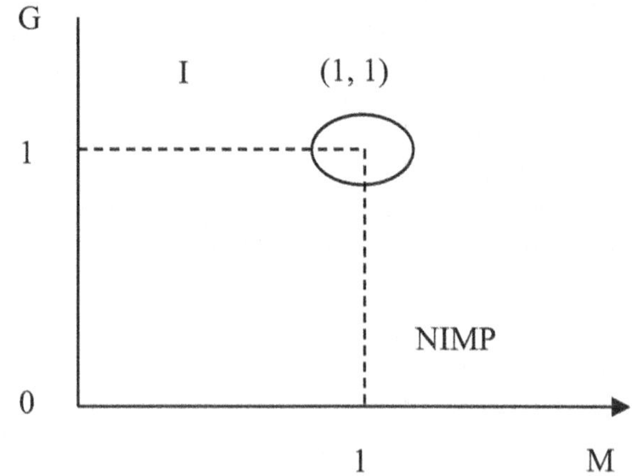

Fig. 3.2 Equilibrium pair of game stage 1

Stage 2

Once the government sets a higher penalty for the manufacturer's non-implementation behavior, the manufacturer can reconsider the payoff. As long as the Eqs. (3.3) and (3.4) are both satisfied, the first game is developed as a mixed-strategy game scenario, in which a probabilistic distribution is placed by the choice of strategies.

$$E - C + S > -P \tag{3.3}$$

$$E - C < 0 \tag{3.4}$$

In this case, let U_{IMP} $(0 \leq U_{IMP} \leq 1)$ denote the probability that the manufacturer chooses the action of 'Implementation' (IMP), $U_{NIMP} = 1 - U_{IMP}$ the probability that the manufacturer chooses 'Non-implementation' (NIMP). Similarly, let V_I $(0 \leq V_I \leq 1)$ denote the probability that the government chooses 'Inspection' (I), $V_{NI} = 1 - V_{NI}$ for 'Non-Inspection' (NI). Thus, the expected payoffs for the manufacturer and the government are expressed as E_M (μ, ν), and E_G (μ, ν), respectively.

$$\begin{aligned} E_M(u, v) = U_{IMP} \times [V_I \times (E - C + S) + (1 - V_I) \\ \times (-C + E)] + (1 - U_{IMP}) \times (-V_I \times P) \end{aligned} \tag{3.5}$$

$$\begin{aligned} E_G(u, v) = V_I \times [U_{IMP} \times (R - S) + (1 - U_{IMP}) \times (P - L)] \\ + (1 - V_I) \times [U_{IMP} \times R - (1 - U_{IMP}) \times L] \end{aligned} \tag{3.6}$$

Let $\frac{\partial E_M(u,v)}{U_{IMP}} = 0$, the Eq. (3.5) becomes:

$$V_I = \frac{C - E}{S + P} \tag{3.7}$$

$$V_{NI} = 1 - \frac{C - E}{S + P} \tag{3.8}$$

Similarly, let $\frac{\partial E_G(u,v)}{V_I} = 0$, the Eq. (3.6) is transformed as follows:

$$U_{IMP} = \frac{P}{S + P} \tag{3.9}$$

$$U_{NIMP} = 1 - \frac{P}{S + P} \tag{3.10}$$

From the Eqs. (3.3), (3.4) and (3.7), $0 < V_I = \frac{C-E}{S+P} < 1$ can be obtained. When $S + P$ is fixed, V_I evolves as an increasing function of C-E, indicating that the manufacturer has less economic return from the implementation of cleaner production.

However, the government needs to improve the efficiency of surveillance to force the manufacturer to take more environmental responsibility. When $C - E$ is fixed, V_I approaches to 0 with $S + P$ increasing, indicating that the government may choose to reduce the probability of inspection gradually. This is because the manufacturer may be influenced either by governmental incentives or heavy fines, thus tending towards cleaner production. From the Eq. (3.9), it is clear that U_{IMP} will increase gradually from 0 to 1, as P rises. In this case, the manufacturer is more willing to use clean technology to avoid the heavy economic penalties for the environmentally unfriendly actions.

Stage 3

If the payoff that the manufacturer chooses implementation (IMP) is higher than that of Non-implementation (NIMP), the following inequalities hold:

$$E - C + S > -P \tag{3.11}$$

$$E - C > 0 \tag{3.12}$$

$E - C > 0$, i.e. $E > C$, indicates that the manufacturer's expected economic return can cover the cost of applying clean technology. As $R > R - S$, the dominant strategy appears to be changed as the manufacturer attempts to comply with cleaner production voluntarily (IMP), while the government no longer needs to supervise the manufacturing (NI), shown in Fig. 3.3.

Based on the above analysis, it is evident that the strategic actions of the government and the manufacturer are strongly determined by the following factors, i.e. the economic penalties (P), government subsidies (S), implementation costs (C) and expected economic returns (E), as shown in Fig. 3.4. In addition, the variance of these game parameters generates three game stages that are involved in the pure and mixed strategy game scenarios. Table 3.2 shows how these factors are measured.

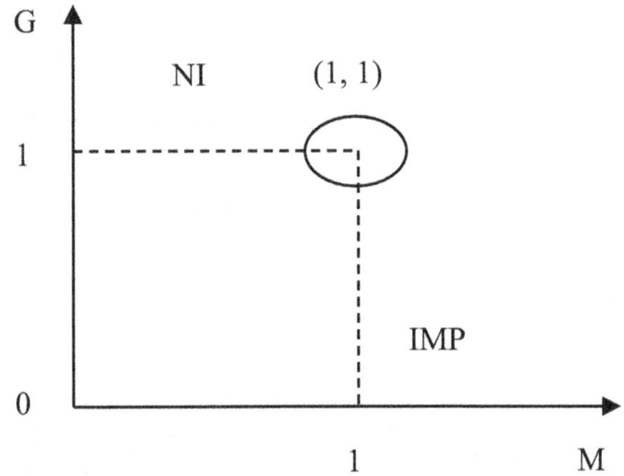

Fig. 3.3 Equilibrium pair of game stage 3

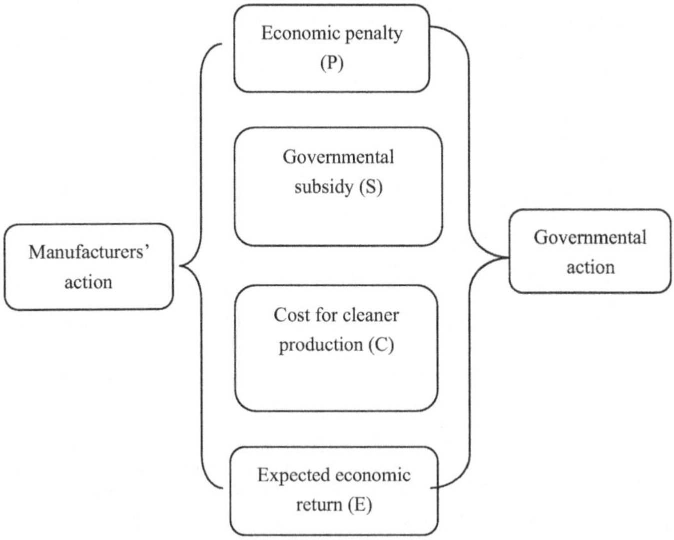

Fig. 3.4 Key influencing factors in the game

3.2.3 An Illustrative Case

A tombarthite manufacturer in Sichuan Province, China, has been studied to access the possible strategic behaviors in a numerical game framework. The tombarthite manufacturer has provided various products, such as praseodymium oxide, lanthana, lanthanum chloride, etc., which comply with the International Standardization Organization (ISO) 9001 and are accredited by the National Qualification Standard. There are two optional technologies for upgrading the existing manufacturing process, i.e. optimization of hydrochloric acid leaching and reconstruction of coiler of the evaporation and concentration tank. For the former technology, the annual consumption of the hydrochloric acid, ammonium hydroxide, ammonium bicarbonate and quick lime is expected to be reduced by 707.5 tonnes, 1387.8 tonnes, 960.6 tonnes, and 1509.3 tonnes, respectively. In addition, it is estimated that 1277 tonnes of fluorine-containing waste water will be reduced annually, while the electricity saving is 3625.8 kwh. For the latter one, there are 51,000 cubic metres of water reduced for consumption annually. Moreover, the numerical values regarding of the two options are shown in Table 3.3.

Analysis of Option 1. As the expected economic return (E) cannot cover the cost of the clean technology ©), i.e. $E < C$, the following two expressions can be derived from the numerical values obtained in Table 3.3.

$$C - E = 187{,}900 \tag{3.13}$$

$$S + P \in [168{,}750,\ 690{,}000] \tag{3.14}$$

Table 3.2 Measurement of the key influencing factors

Factors	Measurement
Economic penalty (P)	The economic penalty is set between 50,000 and 500,000 (CNY). This data is derived from the "Cleaner Production law of Peoples' Republic of China" [8]
Governmental subsidy (S)	Governmental subsidy (S) has been divided into equipment subsidy r and tax break k $S = C * (r + k)$ Equipment subsidy accounts for 15% to 30% of the total investment for cleaner production, and tax break has a proportion of 10% [7, 58]
Cost for cleaner production ©)	The cost ©) is deemed as the initial outlay of investment for cleaner production, composed of cost for technological innovation C_{TI}, equipment upgrading C_{EU}, materials selection C_{MS} and labour training C_{LT}, which can be expressed as followed $C = C_{TI} + C_{EU} + C_{MS} + C_{LT}$
Expected economic return (E)	Economic Return of cleaner production (E) is regarded as the present value of all future benefits which can be expressed as followed [74] $S = \sum_{t=1}^{n} \frac{O_{EA}}{(1+i)^t}$ where O_{EA} denotes expected annual net cash flow from cleaner production i denotes the discount rate annually. Here, i equals to 0.06, derived from the latest policy set by Sichuan Provincial Government [24] O_{EA} can be measured as followed $O_{EA} = N_A + \frac{C}{Y}$ where N_A is annual net profit C is initial outlay of investment Y is depreciable period, assumed as 10 years [24]

Table 3.3 Numerical values of two alternative options for the tombarthite manufacturer

Game parameters China Yuan (CNY)	Option 1 Reconstruction of coiler of the evaporation and concentration tank	Option 2 Optimization of hydrochloric acid leaching
Economic penalty (P)	50,000–500,000	50,000–500,000
Governmental subsidy (S)	118,750–190,000	103,250–165,200
Cost for clean technology ©)	475,000	413,000
Expected economic return (E)	287,100	4,755,700

Based on Eqs. (3.13) and (3.14), $S + P$ can be divided into two mathematical intervals. For $S + P \in [168{,}750, 187{,}900)$, $C - E > S + P$ can be obtained, which has verified the first stage of the pure strategy scenario. Due to the lower economic return, the tombarthite manufacturer is reluctant to implement cleaner production (NIMP). However, the local government should strengthen the inspection (I) by giving specifications of incentive and punitive policies.

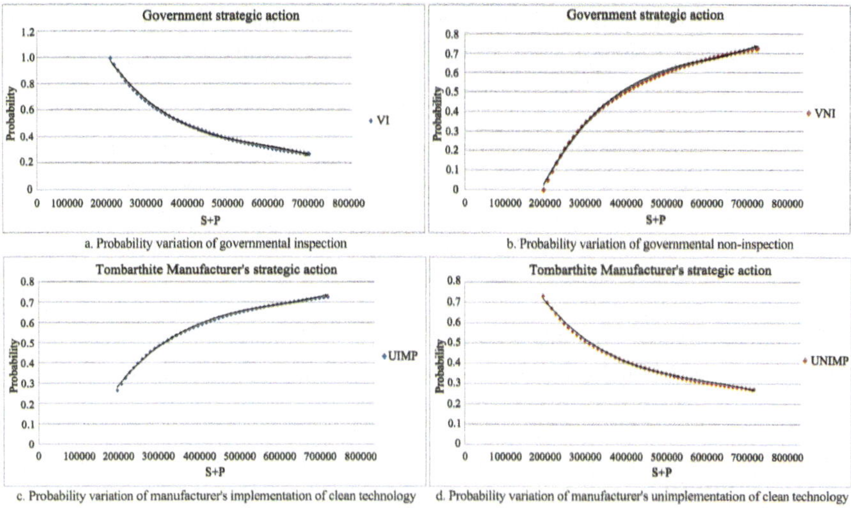

Fig. 3.5 Strategic actions in the mixed-game scenario

For $S + P \in [187{,}900, 690{,}000]$, $C - E > S + P$ can be obtained, which corresponds to the second stage of the game. As $C - E$ is fixed, the probability of inspection (I) of the local government gradually decreases from 1 to 0, as $S + P$ increases, especially as the economic penalty increases, shown in Fig. 3.5a. Oppositely, the probability of Non-Inspection (NI) of the local government increases rapidly as the governmental subsidies (S) increase, shown in Fig. 3.5b. Subsidies can be considered as a form of cost reduction, which help the manufacturer act to be more environmentally friendly [34]. Thus, the tombarthite manufacturer's implementation probability increases with the economic penalty set by the local government, and vice versa, as shown in Fig. 3.5c and d.

Analysis of Option 2. It is apparent that the economic return (E) is considerably greater than the cost of the cleaner production ©), i.e. $E >> C$, which has verified the third stage of the game. Thus, the action of 'IMP' is the dominant stratagem of the tombarthite manufacturer, which implies that more economic and environmental opportunities will be generated, e.g. improvement of product quality, reduction of energy consumption, etc. However, neither government subsidies (S) nor economic sanction (P) plays a key role in the strategic choice, no matter how the parameter changes in this scenario. In this case, it is unnecessary for the local government to intervene strongly in the manufacturing process, indicated as 'NI', due to the s voluntary implementation by the manufacturer. This scenario suggests the best response of interaction between the government and the manufacturer, which can be deemed as a 'win–win' strategy for both parties.

While increasing the economic penalties, the probability of implementation increases for the manufacturer, but the intensity of government inspection decreases. The manufacturer is more willing to apply the clean technology voluntarily to avoid the heavy economic penalties. For the government, it is suggested to gradually reduce the inspection intensity considering the regulatory cost saving. The increase of government subsidy may induce the manufacturer to apply the clean technology. These can be verified by analogy with Dong et al. [12] of the application of game theory to the cleaner production in the Chinese electroplating industry, which indicates that both economic penalties and subsidies have an impact on the policy implementation. Compared with Option 1, it is suggested that the Option 2, i.e. 'optimization of hydrochloric acid leaching', should be more applicable to the tombarthite manufacturer, due to the greater economic return, energy saving, reduction of environmental pollution etc. Only if the government subsidy can be set as high enough, it is expected that the manufacturer will choose the reconstruction of coiler of the evaporation and concentration tank to improve the existing process.

3.2.4 Summary

The application of game theory provides a useful insight to better understand the interaction between government and manufacturer in the context of a cleaner production game. A "Two-person Non-cooperative" static game framework between manufacturer and government is constructed to reflect the variation trend of strategic actions. The game theoretic analysis aids strategic decision-making by selecting appropriate strategies for both organizations and policy makers to reduce environmental impacts while maximizing the economic benefits.

The game is strongly determined by economic penalties (P), governmental subsidies (S), costs ©) and the expected economic returns (E). A case example is given to help a tombarthite manufacturer select the optional clean technology to improve the existing production process, which provides a useful insight into the application of game theory. The game theoretic result suggests a "win–win" strategic situation as the best interaction, which indicates that the manufacturer implements the clean technology voluntarily and the government no longer needs to intervene intensely in the manufacturer's environmentally unfriendly behavior.

3.3 A Game Theoretical Simulation

This section presents the possible game scenarios in the context of cleaner production, generated by a "Two-person Non-cooperative" game model, which explores different strategic options chosen by the government and the manufacturers, to promote cleaner production while maximizing the economic and environmental benefits. In addition, the software package 'Gambit' is selected for the analysis to provide a more visual representation of the game problem situation. By using the 'Gambit' tool for game analysis, the results can be more easily applied to aid decision making in the development of cleaner production. This application will not only help policy makers e.g. government, to better understand the dilemma of cleaner production and the different interactions resulting from the game theoretical analysis, but also help the manufacturers to identify mitigation strategies to act in accordance with the principles of cleaner production.

3.3.1 Model Formulation

In its simplest form, a "Two-Person Non-Cooperative" game model is built with two players: Government (G) and Manufacturers (M), respectively. A game is defined as non-cooperative if there is no pre- communication or agreement, i.e. the strategies chosen by the players cannot be coordinated in such a game situation [6, 19]. This model is based on the consideration of 'fiscal measures' by the government and 'cost–benefit analysis' by the manufacturers, respectively. Assume that each has two available strategies to promote cleaner production. For the government, the strategies are: {R, NR}, which indicate whether the government should regulate ®) the manufacturing process and the market competition in order to promote cleaner production or not regulate (NR), while maintaining the current relationship between supply and market demand. Correspondingly, there are two strategies for the manufacturers, which are {CH, NCH}. Action CH means that the manufacturers decide to change their existing process technology or equipment to achieve better product quality and market expansion, while NCH means no change in existing business practices. Table 3.4 shows the expected payoffs for the two players with different actions.

In the payoff matrix, T_B corresponds to the government subsidy in the form of a tax break to help manufacturers implement cleaner production; P denotes the economic penalty if the manufacturers still insist on environmentally unfriendly production; and C represents the cost of technological innovation by the manufacturers for cleaner production. E is the economic benefit to the government in terms of revenue if the manufacturers improve production technology while minimizing environmental impacts, such as reducing environmental risk and carbon footprint, etc. S is the additional sales that the manufacturers can expect from adopting cleaner production instead of the existing production mode. I is the possible loss resulting from environmental impacts (such as pollution incidents). It is assumed in this study that manufacturers cannot predict the government's actions in the early stages of the game or before the game is played. However, they may pay more attention to the potential economic benefits when deciding whether or not to change. Meanwhile,

Table 3.4 Payoffs matrix between government (G) and manufacturers (M)

G	M	
	CH	NCH
R	$E - T_B, T_B - C + S$	$P - I, - P$
NR	$E, - C + S$	$- I, 0$

the government cannot decide whether to regulate the market initially or not, because the actions of the manufacturers may change or may not change their existing production mode. Thus, their actions are triggered simultaneously, which can be considered static [20]. For any non-cooperative game, the solution is derived from the Nash Equilibrium, which addresses that each player's chosen strategy is a best response to other player's strategies [18, 21]. How the Nash Equilibrium is derived using a Gambit simulation is described in more detail in the following section.

3.3.2 Game Simulation Results

In this section, the simulation software tool 'Gambit', integrated with an Excel spreadsheet, is used to express the dilemma and possible game actions between the government and the manufacturers. Gambit is a free open source software for game theory simulation that allows users to customize its features for their own needs . In addition, the game solution is encapsulated in a user-friendly interface where users can easily invoke the functions to simulate the game situation [38].

According to the basic game model in Table 3.4, three possible game scenarios are provided in order to select different pairs of strategies. These game scenarios are generated by inserting different numerical values in the form of a logical unit. Here, the numerical values for each parameter are generated by the random number function of Excel spreadsheet. The details of the game scenario analysis are described in the following sections.

Game scenario one

Game scenario one illustrates the preliminary stage of the game situation, which reflects the difficulty of developing a cleaner production program. For the government, the supervision and regulation of cleaner production is still insufficient with ambiguous policies of sanctions and incentives. Furthermore, any new change may pose an economic risk to the manufacturer. For instance, if the additional sales of environmentally friendly products cannot initially cover the total economic costs, the manufacturers will be reluctant to implement cleaner production. Thus, once $S - C + T_B < -P$, manufacturers will prefer to pay the penalties for environmentally unfriendly production. Selected model parameters for this game simulation are presented in Table 3.5. In order for wider application, all the parameters are expressed in terms of a logical unit. For example, costs can be indicated as 1 unit to represent the real monetary value. The tax break can be provided in terms of 'Capital allowance', currently around 22% in the UK for plant and machinery, and thus it is suggested in this model that 22% of the technological innovation cost can be saved by upgrading the existing machinery, equipment or tools to promote cleaner production. The factors on revenue and additional sales are both set as random variables, which are generated by the random function of Excel, 10 to 10,000 times more than the cost. In this scenario, we assume that additional sales can cover at most 50% of the cost (0.1 to 0.5). Moreover, the penalty factor related to economic sanctions must be set at a very low level (0 to 0.2), which is estimated by the equation $S - C + T_B < -P$, as well as the value range of the additional sales factor (S), shown in Table 3.5. Otherwise, if the penalty for manufacturers is set at a higher level, the manufacturers would mostly opt to change their production process. Thus, the dominant strategy pair will be ®, NCH) in this scenario.

Once the model parameters have been determined by Table 3.5, the Nash Equilibrium is stable and will not change no matter how many times the game model is run. Thus, a random set of numerical values has been selected to represent the game situation using the 'Gambit' tool (see Fig. 3.6a). In the Gambit tool, the dominated strategies can be eliminated iteratively using the dominance panel. The 'strictly dominated' action is the opposite of the dominant strategy, which is defined as "always worse than another, regardless of beliefs at the information set" [39]. In this scenario, CH is the strictly weak strategy for the manufacturers to eliminate first (see Fig. 3.6b), because its expected payoffs are the lowest compared to choosing NCH. Accordingly, if the manufacturers determine not to change their current production mode, non-regulation (NR) becomes the dominant strategy. Figure 3.6c shows the second round of dominated strategy elimination, with 'NR' crossed out. Once all the dominated strategies have been eliminated, the dominant strategy pair emerges. In scenario one, the result in Fig. 3.6d suggests that the government should regulate the manufacturing process by choosing 'R', while the position of the manufacturers remains unchanged ('NCH'). This is because the economic sanctions are insufficient in this case.

Table 3.5 Model parameters of game scenario one

Parameters	Conversion factors	Note
Cost ©)	1 unit	Rand is a function by which the random number from 0 to 1 will be selected Revenue factor is between 10 and 10,000 Sales factor is between 0.1 and 0.5 Fine factor is between 0 and 0.2 Environmental loss factor is 10,000
Tax break (T_B)	Cost × 22%	
Economic benefit (E)	Cost × rand × revenue factor	
Additional sales (S)	Cost × rand × sales factor	
Economic sanction (P)	Cost × rand × fine factor	
Environmental loss (I)	Cost × rand × loss factor	

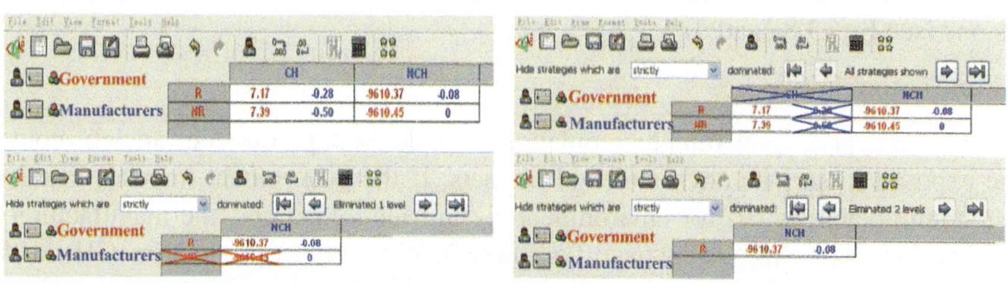

Fig. 3.6 Gambit simulation for game scenario one

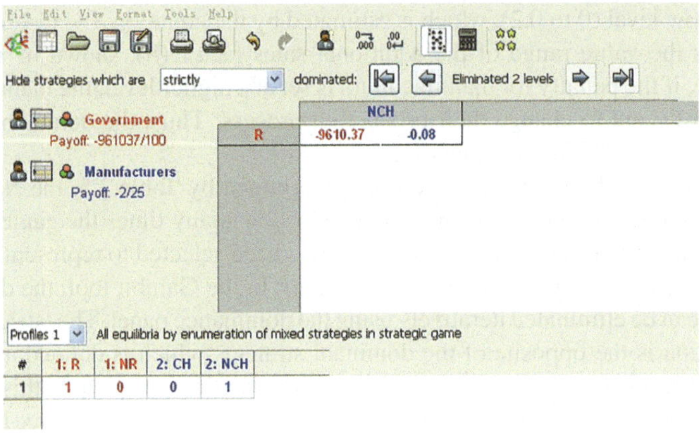

Fig. 3.7 Computation result of nash equilibrium in game scenario one

The remaining strategy pair (R-NCH) represents the unique and stable Nash Equilibrium for this game solution. Moreover, this is a pure strategy game, i.e. a game in which the chosen strategies have a probability of one and the others zero [46]. Fig. 3.7 shows that the probability equals one if the government chooses 'R' and the manufacturers choose 'NCH'.

The traditional method of computing Nash Equilibrium is strongly deterministic. However, the 'Gambit' tool provides a novel approach called 'Quantal Response Equilibrium' (QRE) to search for the subset of Nash equilibria 'imperfect' or 'noisy' expectations logic instead of the perfectly rational expectations logic [37]. Figure 3.8 demonstrates how the 'QRE' approach is used. The QRE model is a function of probability distribution and the error of choice selection, Lambda (λ), which is related to the level of error in expectations. For $\lambda = 0$, the actions are composed of errors, and for $\lambda \to \infty$, there is no error. If both players, the government and the manufacturers, evaluate their expected payoffs in an unbiased way, the probabilities of choosing each strategy should beequal at the beginning of the game, and thus all graphs in Fig. 3.8 start at 0.5. As Lambda tends to infinity (∞), the Nash Equilibrium converges to a unique value. For example, in this game scenario, the strategy pair that represents the Nash Equilibrium is 'R-NCH', and the probability of choosing both strategic actions is one (see Fig. 3.8a and d).

Game scenario two

Without effective government regulation of industrial production, society could be exposed to serious environmental impacts. If the government imposes heavier economic penalties (P) for environmentally unfriendly actions, the manufacturers will be forced to compare the payoff of implementing cleaner production (CH), with that of maintaining existing methods (NCH) in order to identify the most beneficial strategy. Game scenario two suggests that if $S - C + T_B > -P$, there is no dominant strategy in the game. Thus, a 'mixed-strategy' game scenario is generated to find the

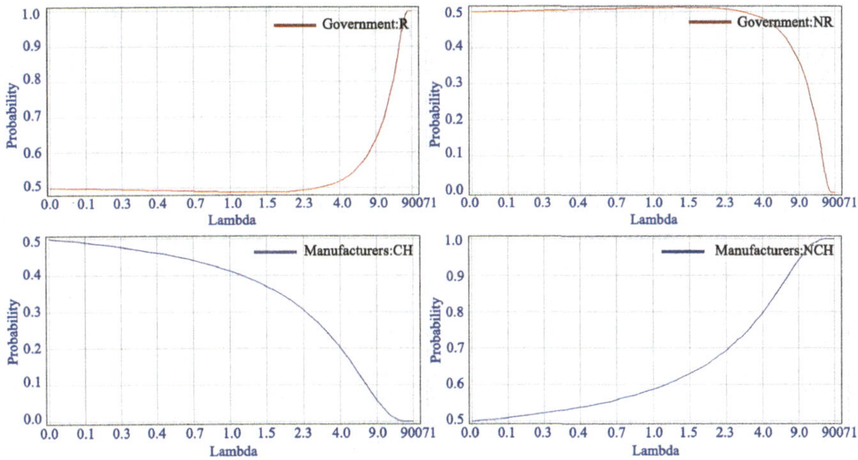

Fig. 3.8 Quantal response equilibrium approach for game scenario one

corresponding Nash Equilibrium. In a pure strategy game, as in game scenario one, each player can only choose one specific strategy. A mixed-strategy equilibrium, in contrast to a pure strategy equilibrium, can be defined as "at least one player involved in the game will place a probability distribution for the alternative strategies" [46]. Table 3.6 presents the parameters selected for the simulation of the mixed strategy game of the second scenario. In this case, the level of economic penalties (P) for environmentally unfriendly actions is now set to severe, i.e. weighted as 10,000 times higher than the cost.

Figure 3.9 shows the variation of Nash Equilibria computed by the 'Gambit' tool, with the sales factor being varied from 0.1 to 1. It can be found that the probability that the government chooses non-regulation (NR) while the manufacturers choose change (CH) is very close to unity. Clearly, this strategic choice pair (NR-CH) is strongly determined by the government penalty. Only if the penalty is high enough, the manufacturers will consider changing their current industrial process by upgrading the existing equipment or technology for cleaner production.

Table 3.6 Parameters of game scenario two

Parameters	Conversion factors	Note
Cost ©)	1 unit	Rand is a function by which the random number from 0 to 1 will be selected Revenue factor is between 10 and 10,000 Sales factor is between 0.1 and 1 Fine factor is 10,000 Environmental loss factor is 10,000
Tax break (T_B)	Cost × 22%	
Economic benefit (E)	Cost × rand × revenue factor	
Additional sales (S)	Cost × rand × sales factor	
Economic sanction (P)	Cost × rand × fine factor	
Environmental loss (I)	Cost × rand × loss factor	

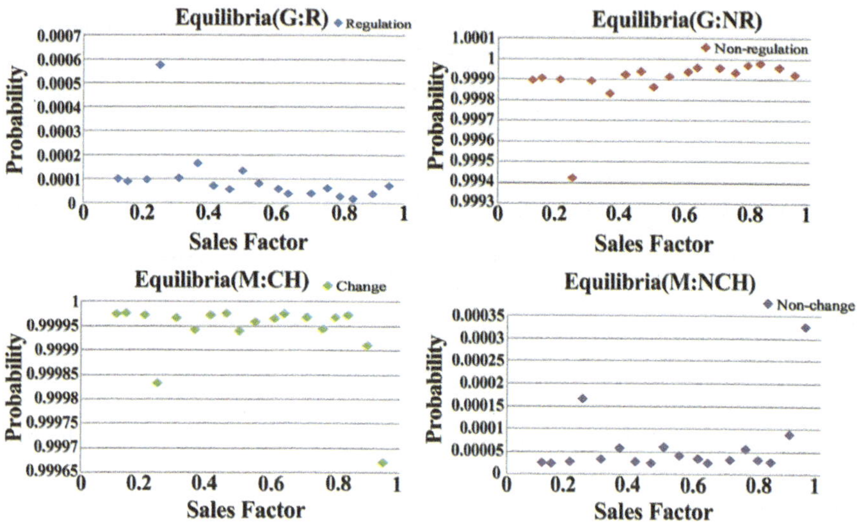

Fig. 3.9 Nash equlibria of game scenario two

Game scenario three

As "green consumerism" becomes more influential in the marketplace, cleaner production will develop to be matured correspondingly, and the manufacturers will increasingly seek to benefit from the business opportunities connected with selling 'products' that are perceived as environmentally friendly. Once the additional sales fully cover the total cost of adopting cleaner production methods ($S > C$), the government will no longer need to take action to regulate the process, and the manufacturers will choose to develop cleaner production voluntarily. This game scenario suggests that the dominant strategy is for the government to choose 'NR' and for manufacturers to choose 'CH', which is the optimal outcome among the three game situations. Table 3.7 summarizes the conversion factors used to determine the model parameters.

As long as 'additional sales' (S) is above costs (C), 'NR-CH' (Non-regulation for the government, change for the manufacturers) is the unique Nash Equilibrium of

Table 3.7 Model parameters of game scenario three

Parameters	Conversion factors	Note
Cost ©)	1 unit	Rand is a function by which the random number from 0 to 1 will be selected Revenue factor is between 10 and 10,000 Sales factor is between 10 and 10,000 Fine factor is 10,000 Environmental loss factor is 10,000
Tax break (T_B)	Cost × 22%	
Economic benefit (E)	Cost × rand × revenue factor	
Additional sales (S)	Cost × (1 + rand × sales factor)	
Economic sanction (P)	Cost × rand × fine factor	
Environmental Loss (I)	Cost × Rand × Loss factor	

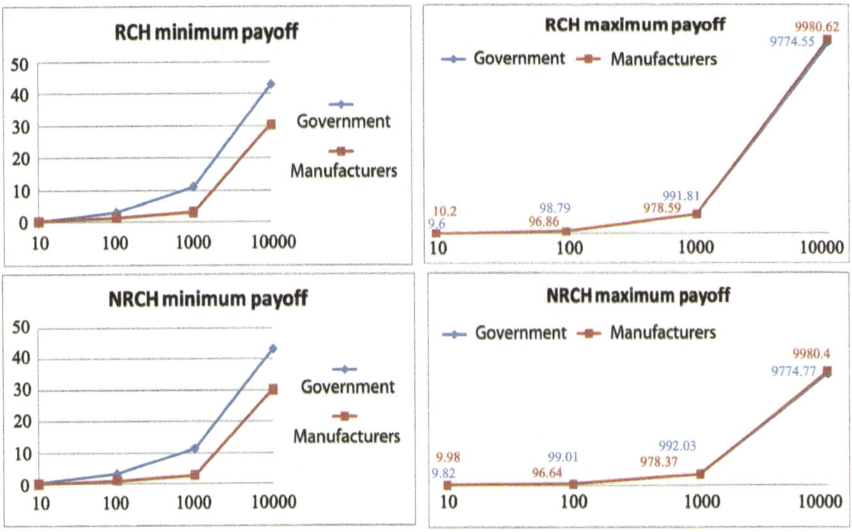

Fig. 3.10 Maximum and minimum payoffs with different factors

this game scenario no matter how the factors change. Figure 3.10 shows the possible maximum and minimum payoffs of strategies of the 'R-CH' and 'NR-CH' as both the revenue and sales factors are varied from 10 to 10,000.

As long as the inequality $S > C$ is satisfied, the Nash Equilibrium will not change in this scenario. Figure 3.11 shows a random set of numerical values being selected to simulate the game situation using the Gambit tool (assuming 10 for both revenue and sales factors in this example). the analysis of game scenario one, the dominated strategy and its corresponding payoff should be eliminated. Fig. 3.11b shows that no matter whether the government chooses to regulate or not to regulate, 'NCH' is the weak strategy for the manufacturers that should be eliminated first. If the action of 'NCH' is eliminated in the first round, 'R' is the dominated strategy for the government in the second turn, while 'CH' is choosen by the manufacturer chosen Fig. 3.11c). Once all the dominated strategies have been eliminated, Fig. 3.11d suggests that the dominant strategy is that 'NR' to be choosen by the government and 'CH' by the manufacturers. This result also suggests a pure strategy pair with a probability of one being calculated for both strategies 'NR' and 'CH', respectively (see Fig. 3.12).

As in the first game scenario, the 'QRE' approach can also be applied to this scenario in order to find the unique Nash equilibrium. Figure 3.12 shows that the strategic action pair 'NR-CH' is the unique and stable solution of the Nash Equilibrium found by the 'Gambit' tool.

In contrast, the 'QRE' curves shown in Fig. 3.13, which present the steady trend of inceasing or decreasing probability, first fluctuate and then stabilize. For example, if the government chooses the action of 'R', the probability firstly rises from 0.5 to over

Fig. 3.11 Gambit simulation for game scenario three

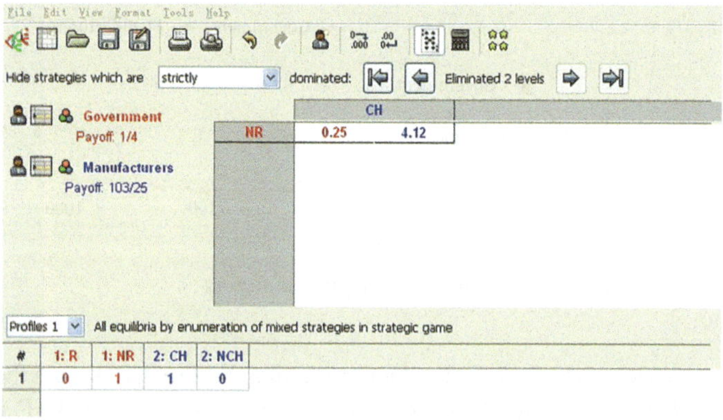

Fig. 3.12 Computation result of Nash equilibrium in game scenario three

0.6, then falls backand reaches 0 (see Fig. 3.13a). This phenomenon results from the principle of maximum utility followed by different actions of the government, ‘R’ and ‘NR’ respectively. From the payoff matrix in the game scenario three, it can be seen in Fig. 3.11 that the government can obtain the maximum benefit by choosing ‘R’ to the extent of 3321.95 logic unit. That is why the probability of ‘R’ shows an upward trend from 0.5 to 0.6 in Fig. 3.13a. However, taking the possible actions of the manufacturers into account, e.g. if ‘CH’ is chosen, the individual rational behaviour would lead the government to reconsider the payoffs. By contrast, the government can obtain more benefit to the extent of 0.25 logic unit if ‘NR’ is chosen. Thus, the probability of choosing ‘R’ falls back to 0.5, and then gradually reduces to 0.

According to the above game simulation, it is apparent that the strategic actions that the government and manufacturers will take are mainly determined by the following factors: Economic penalty (P), Technological innovation cost (C) and Additional sales (S). Figure 3.14 illustrates the interaction between the influencing factors and the strategic actions.

It is clear that the economic penalty (P) set by government policy must be high enough to force the manufacturers to change their current production mode actively, thus implementing cleaner production and providing ‘environmentally friendly’

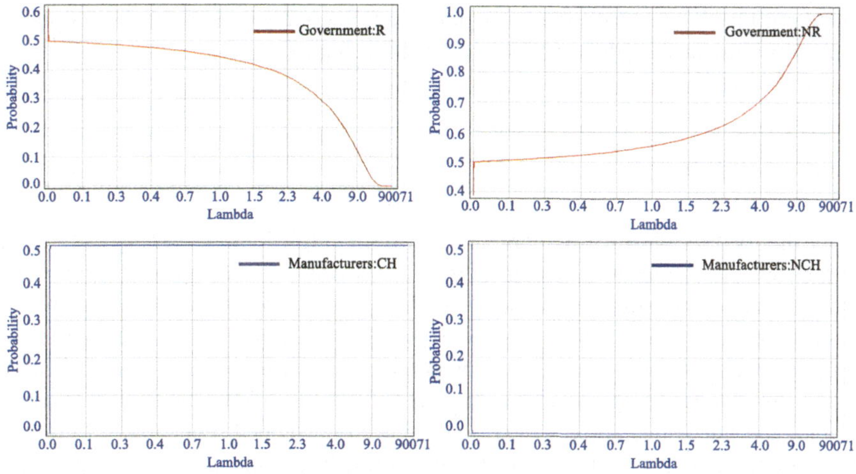

Fig. 3.13 Quantal response equilibrium approach for game scenario three

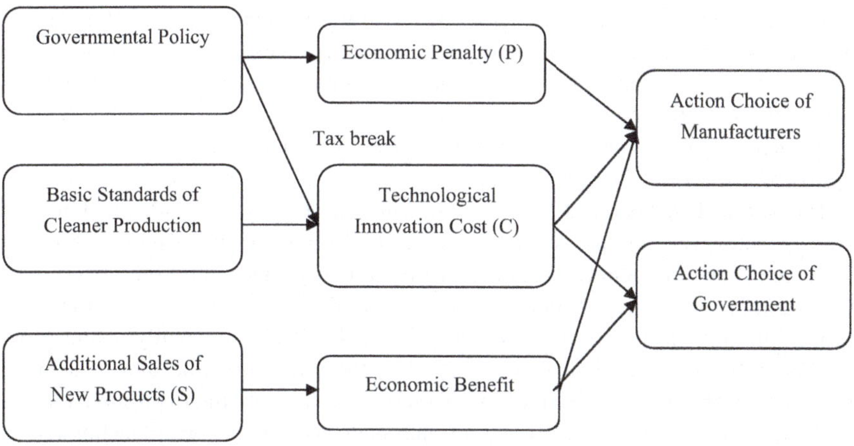

Fig. 3.14 Influencing factors and strategic action

products. The technological innovation cost ©) should be in accordance with the basic standards or principles of cleaner production, to ensure that new products are associated with reduced environmental risk and carbon footprint, and are sustainable. For food, and those materials with which people will have direct contact, the criteria of production should be much more stringent. The potential economic benefit to the manufacturers is determined by the additional sales profit (S), which will also give rise to increased government revenue. Moreover, the profit accruing from additional sales is strongly determined by consumer behaviour towards 'environmentally sound products', whereby public acceptance automatically determines the actions of manufacturers, i.e. whether to 'Change' or 'Not change'.

3.3.3 Summary

Game theory offers an effective approach to formulating a dominant strategy designed to improve performance across the three dimensions of the Triple Bottom Line (Economic, Society and Environment). This study has applied game theory to better understand the dilemmas that may exist between government and manufacturers, and to provide insightful perspectives on potential problems in the context of cleaner production. By assigning different numerical values to the model parameters in terms of a logical unit, three game scenarios have been developed sequentially to reflect the possible changes in the actions of government and manufacturers that could accompany the implementation of cleaner production.

The game model mainly depends upon the variation of three important factors: innovation cost ©), economic penalty (P) and additional sales of 'environmentally friendly products' (S). Game scenario one illustrates the consequences of a weak or ambiguous sanction policy. While the government may seek to regulate the process in the early stages of cleaner production, if financial penalties are insufficient the dominant strategy for the manufacturers is not to change existing production practices. It is assumed in game scenario two that the government recognies the importance of regulation and supervision, and increases the penalties accordingly.

The game then becomes a two person mix-strategy game with no dominant strategy. However, if the penalties are set high enough, the manufacturers opt to change their production technology to be much 'cleaner', and the government will gradually adjust its regulatory policy. Game scenario three illustrates that once cleaner production has been implemented, and the 'environmentally sound products' have become the mainstream of market, the government no longer needs to intervene so strongly, because the manufacturers are then willing to promote cleaner production to obtain higher profits, additional sales to cover the total of technological innovation, equipment upgrading, and so on. Thus, the dominant strategy ultimately becomes non-regulation for government and change for manufacturers ('NR-CH').

3.4 Game Theoretical Analysis to Green Supply Chain

This section proposes a game theory approach that models the likely behaviour of manufacturers in response to environmental risk and carbon emissions reduction in the context of the green supply chain. Although the concept of green supply chain management has emphasized increasing economic returns by reducing the environmental impacts [73, 53], there has been limited discussion of the extent to which the environmental risk and carbon emissions reduction should be incorporated into the management process.

A previous study indicated how life cycle carbon emissions could be presented as an aid to customer choice, and thus as a means of sending an environmental signal to manufacturers [66]. This study extends the analysis to consider the perspective of manufacturers. Using game theory, with a novel underlying mathematical model, this study identifies the most appropriate strategy in the context of "green supply chain management", for manufacturers of commercial products to optimize the three dimensions of sustainability. This study proposes game theoretical analysis for strategy selection to reduce environmental risk and lifecycle carbon emissions by manufacturers, initially without government supervision, and subsequently, with government supervision.

3.4.1 Model Formulation

As discussed above, the environmental risk needs to be controlled and managed at an early stage of product development. Furthermore, in order to understand what the level of risk should be and to assess whether more should be done to reduce the risk, it is necessary to establish certain risk criteria to determine the level of acceptability[3]. Thus, the principle of "tolerability of risk" provides a framework for judging whether society should accept the risk [25]. Importantly, this framework provides a benchmark for a tolerable risk, see Fig. 3.15, which cannot be considered broadly acceptable without the implemention of risk control measures, and also fills the gap between what is absolutely safe and what is absolutely unsafe, i.e. socially intolerable [9]. Between the intolerable region and the broadly acceptable region, it is necessary to demonstrate that the risk is 'as low as reasonably practicable' (ALARP), i.e. that risk mitigation is in place. Within the ALARP region, a decision can be made whether to reduce the level of risk further or to accept the existing level of risk based on the level of the risk and the costs associated with controlling the risk [17]. The principle is well established in some heavy industries, such as nuclear, chemical, etc. The HSE has also proposed notional boundaries for each region (see Fig. 3.15) in terms of the risk of death or serious injury to a member of the public. A risk of 1 in 10^4 per year to any member of the public is the maximum that should be tolerated from any large industrial plant in any industry. The level of broadly acceptable risk is suggested to be 1 in 10^6 per year [25].

As there is no specific standard for carbon emissions in EU countries, this study suggests that, by analogy, the boundary between the intolerable and tolerable regions should be represented by the 2020 carbon emission target, and the boundary between the tolerable and broadly acceptable regions should be represented by the reduction target at the year 2050 (see Fig. 3.15). According to the UK Climate Change Act 2008 [55], total carbon emissions are expected to be reduced to, at least, 26% below the 1990 baseline by 2020 and 80% by 2050, respectively. The 1990 baseline represents the total UK net emissions of the identified greenhouse gases for that year, e.g.

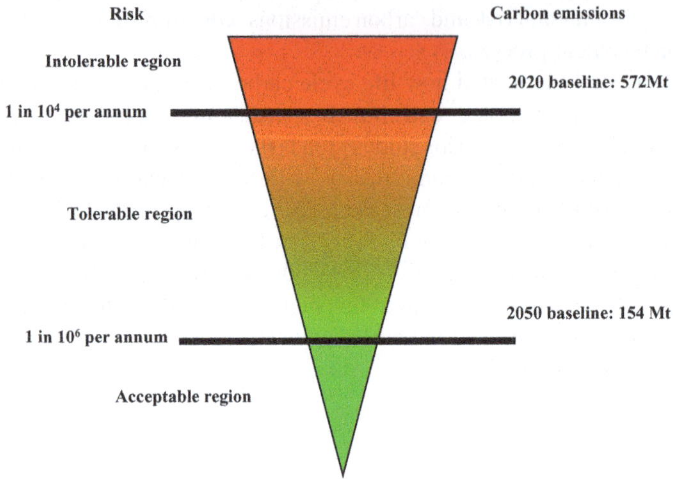

Fig. 3.15 Intolerable, tolerable and acceptable regions of risk and associated analogous carbon emissions baseline

CO_2, CH_4, N_2O, HFCs, PFCs and SF_6 etc. [43]. The total aggregate anthropogenic emissions of greenhouse gases, excluding emissions from land use, land-use change and forestry, was approximately 774Mt for the UK in 1990 [56]. Thus, the carbon emissions for the years 2020 and 2050 can be calculated from the 1990 baseline as approximately 572 Mt and 154 Mt, respectively. Moreover, we assume that these emissions can be seen as the total emissions contributed by the national supply chain network.

Notations

PF	Payoff of the conventional supply chain
PF'	Payoff of the supply chain in the context of Green Supply chain management
cf_i (i = 1,2,3)	Carbon emissions of the supply chain at different level; cf_1 represents the carbon emissions at the acceptable level; cf_2 represents the carbon emission at the tolerable levels; cf_3 represents the carbon emissions at the intolerable level
R_i (i = 1,2,3)	Inherent risk of the supply chain containing 'hazardous materials' at different levels; R_1 represents the risk at the acceptable level; R_2 represents the risk at the tolerable level; R_3 represents the risk at the intolerable level
Ct	Carbon tax
Cr	Cost per unit of risk

It is proposed that the manufacturer of 'hazardous' materials in the supply chain be assigned a carbon emissions level cf and an inherent risk R, as well as a value of PF reflecting the payoff of the conventional supply chain without taking the carbon emissions and risk into account. The payoff function PF' in the context of the green supply chain is expressed as follows:

$$PF' = PF - (R \times Cr) - (cf \times Ct) \tag{3.15}$$

In Eq. (3.15), the inherent risk of the 'hazardous materials' supply chain can be measured by multiplying the consequence $N(i)$ by the corresponding probability $P_r(i)$ in a given hazard scenario [42]. The consequence of the hazard in this study is considered to be the number of fatalities resulting from environmental accidents or other adverse events attributed to the supply chain, such as those might occur in the different scenarios (i), process of production, transportation and final disposal.

$$R = P_r(i) \times N(i) \tag{3.16}$$

A common approach measuring the life cycle carbon emissions (cf) is to apply a greenhouse gases emission factor (G_f) multiplied by an activity data factor (H_d) [10].

$$cf = G_f \times H_d \tag{3.17}$$

The activity data factor of a supply chain network mainly reflects the energy consumption in the process of production and transportation, and the annual volume of various waste materials generated. Thus, Eq. (3.17) can be transformed as follows:

$$cf = \sum_i A_{pi} G_{pi} + \sum_j B_{tj} G_{tj} + \sum_k C_{dk} G_{dk} \tag{3.18}$$

where A_{pi} denotes the activities contributed to annual carbon emissions in the process of production, e.g. various energy consumption (burning oil, coal, compressed natural gas etc.). Similarly, B_{tj} indicates the annual transportation distances according to different vehicles, and C_{dk} is the annual volume of various waste materials generation, e.g. paper, plastic, glass etc. G_{pi}, G_{tj} and G_{dk} indicate the corresponding greenhouse emission factors for production, transportation and disposal, separately.

The cost per unit of public risk reduction (Cr) can be derived from the compensation for fatalities. For example, in the UK, the value of a prevented fatality (VPF) associated with a risk reduction of one in one hundred thousand is given as £1,000,000 (2001 prices) [26]. Moreover, as there is no specific standard for the cost of carbon emissions, Ct can be used instead of energy tax in the different scenarios.

If the manufacturers do not consider the risk and carbon emissions reduction in the supply chain, the payoff of the entire supply chain will not change, i.e. PF' is equal to PF. However, if the manufacturers decide to reduce the risk and carbon footprint within the supply chain, extra costs may be incurred for supplier selection, technology improvement, equipment upgrades, and so on. On the other hand, the quality and safety of the products can be improved, which is expected to result in additional sales profits. This will be discussed in the following section on the assumptions made for the game theoretical analysis.

3.4.2 Model Hypothesis

When the current carbon emissions level of the supply chain cf is within the tolerable carbon emissions area (see Fig. 3.16), which is higher than the upper limit f_1 of the acceptable carbon footprint area, but lower than the upper limit f_2 of the tolerable carbon emissions area, the manufacturer has two potential strategic choices: either to reduce or to maintain the carbon emissions at their current level. Once carbon emissions cf is in the intolerable region, i.e. above the f_2 criterion, manufacturers in the supply chain should reduce carbon emissions to the tolerable area at least, if not to the acceptable area. It is assumed that the acceptable level should be the 'optimal region' in terms of green supply chain management to minimize the risk and carbon emissions. In addition, the current risks, r_1 and r_2, should correspond to equivalent carbon emissions, f_1 and f_2, according to the thresholds shown in Fig. 3.16. For the purpose of analysis, we also define the acceptable, tolerable and intolerable regions, as regions 1, 2, and 3, respectively.

It is assumed that additional economic benefits can be generated as a result of the reduction in risk and carbon emissions within the supply chain network, and will thus be welcomed by consumers as 'environmentally friendly'. In Fig. 3.17a, let P_{R1}, P_{R2}, P_{R3} denote the yield gaps, i.e. the differences between the various economic benefits, as a result of the environmental risk decreasing from the tolerable region to the acceptable region, from the intolerable region to the acceptable region and from the intolerable region to the tolerable region, respectively. Similarly, in Fig. 3.17b, let P_{cf1}, P_{cf2} and P_{cf3} denote the revenue gap as a result of the carbon emissions decreasing from the tolerable region to the acceptable region, from the intolerable region to the acceptable region, and from the intolerable region to the tolerable region, respectively.

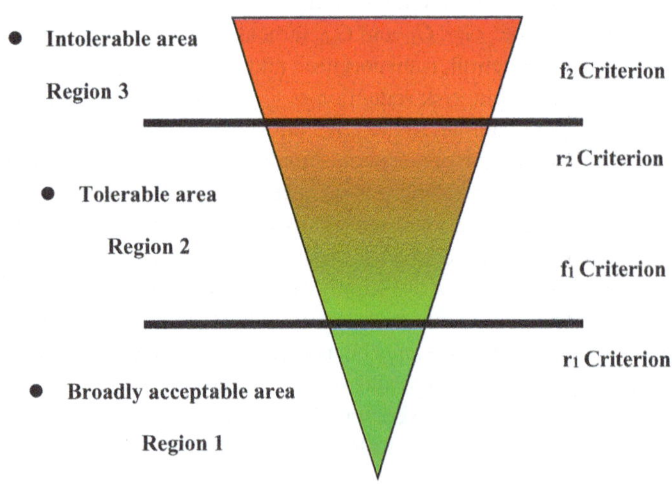

Fig. 3.16 Criteria of risk and carbon emissions reduction

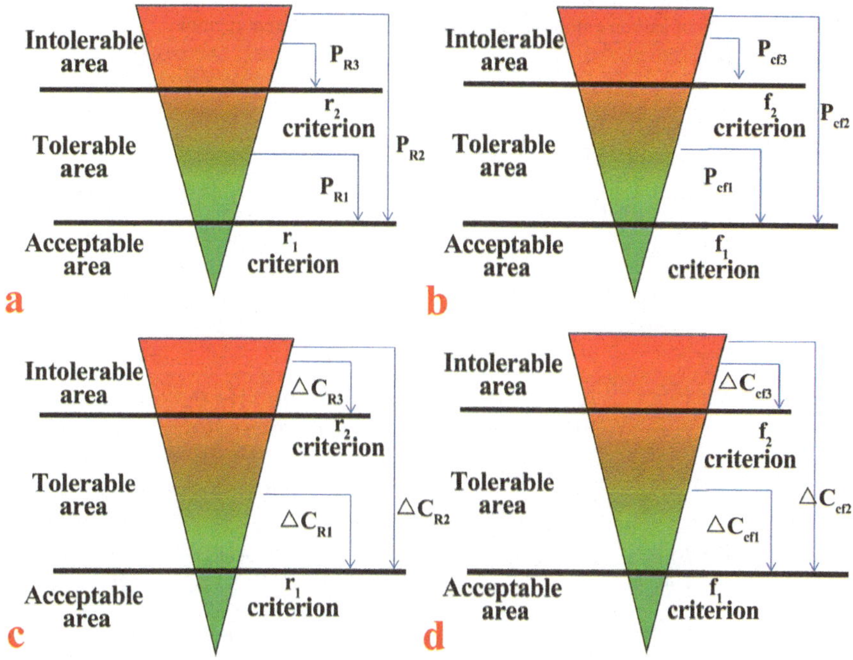

Fig. 3.17 Additional profit and cost in different risk or carbon emissions region

As shown in Fig. 3.17c, let the additional cost of reducing risk from the tolerable level to the acceptable level be ΔC_{R1}, from the intolerable level to the acceptable level be ΔC_{R2} and from the intolerable level to the tolerable level be ΔC_{R3}. Similarly, in Fig. 3.17d, let the additional cost of reducing the carbon footprint from the tolerable to the acceptable level be ΔC_{cf1}, from the intolerable to the acceptable level be ΔC_{cf2}, and from the intolerable to the tolerable level be ΔC_{cf3}.

3.4.3 A Cost–Benefit Game of Manufacturers

In the first scenario, we assume that the strategic actions adopted by the manufacturers within the supply chain are equal and simultaneous. According to the above basic payoff function and assumption, the payoff matrix of manufacturers at different risk or carbon emissions level is given in Table 3.8. Moreover, Table 3.8 can be divided into several subtables for the following analysis of manufacturers' strategic actions to improve the current carbon emissions and risk level, but without government supervision.

When carbon emissions and risk reduction are at the acceptable level, the manufacturers in the supply chain will cease to focus on further carbon emissions or risk reduction, as denoted by "$(PF - R_1 Cr - cf_1 Ct, PF - R_1 Cr - cf_1 Ct)$" in Table 3.8.

Table 3.8 Payoffs of carbon emissions and risk reduction in different regions

Carbon emissions	Risk				
	Acceptable (r_1)	Tolerable (r_2) ALARP (as low as reasonably practicable)		Intolerable	
Acceptable (f_1)	$(PF - R_1C_r - cf_1C_t,$ $PF - R_1C_r - cf_1C_t)$	Maintain	Reduce to acceptable level	Reduce to acceptable level	Reduce to tolerable level
Tolerable (f_2)	Maintain	$(PF - R_2Cr - cf_2Ct,$ $PF - R_2Cr - cf_2Ct)$	$(PF - (R_2 - r_1)Cr - cf_2Ct,$ $PF - (R_2 - r_1)Cr - cf_2Ct - \Delta C_{R_1} + P_{R_1})$	$(PF - cf_2Ct - (R_3 - r_1)Cr,$ $PF - (R_3 - r_1)Cr - cf_2Ct + P_{R_2} - \Delta C_{R_2})$	$(PF - cf_2Ct - (R_3 - r_2)Cr,$ $PF - (R_3 - r_2)Cr - cf_2Ct + P_{R_3} - \Delta C_{R_3})$
	Reduce to acceptable level	$(PF - R_2Cr - (cf_2 - f_1)Ct - \Delta C_{cf_1} + P_{cf_1},$ $PF - R_2Cr - (cf_2 - f_1)Ct)$	$(PF - (R_2 - r_1)Cr - (cf_2 - f_1)Ct - \Delta C_{cf_1} + P_{cf_1},$ $PF - (R_2 - r_1)Cr - (cf_2 - f_1)Ct - \Delta C_{R_1} + P_{R_1})$	$(PF - (R_3 - r_1)Cr - (cf_2 - f_1)Ct - \Delta C_{cf_1} + P_{cf_1},$ $PF - (R_3 - r_1)Cr - (cf_2 - f_1)Ct - \Delta C_{R_2} + P_{R_2})$	$(PF - (R_3 - r_2)Cr - (cf_2 - f_1)Ct - \Delta C_{cf_1} + P_{cf_1},$ $PF - (R_3 - r_2)Cr - (cf_2 - f_1)Ct - \Delta C_{R_3} + P_{R_3})$
Intolerable	Reduce to acceptable level	$(PF - \Delta C_{cf_2} + P_{cf_2} - (cf_3 - f_1)Ct - R_2Cr,$ $PF - R_2Cr - (cf_3 - f_1)Ct)$	$(PF - \Delta C_{cf_2} + P_{cf_2} - (cf_3 - f_1)Ct - (R_2 - r_1)Cr,$ $PF - (R_2 - r_1)Cr - (cf_3 - f_1)Ct + P_{R_1} - \Delta C_{R_1})$	$(PF - \Delta C_{cf_2} + P_{cf_2} - (cf_3 - f_1)Ct - (R_3 - r_1)Cr,$ $PF - (R_3 - r_1)Cr - (cf_3 - f_1)Ct + P_{R_2} - \Delta C_{R_2})$	$(PF - \Delta C_{cf_2} + P_{cf_2} - (cf_3 - f_1)Ct - (R_3 - r_2)Cr,$ $PF - (R_3 - r_2)Cr - (cf_3 - f_1)Ct + P_{R_3} - \Delta C_{R_3})$
	Reduce to tolerable level	$(PF - \Delta C_{cf_3} + P_{cf_3} - (cf_3 - f_2)Ct - R_2Cr,$ $PF - R_2Cr - (cf_3 - f_2)Ct)$	$(PF - \Delta C_{cf_3} + P_{cf_3} - (cf_3 - f_2)Ct - (R_2 - r_1)C_r,$ $PF - (R_2 - r_1)Cr - (cf_3 - f_2)Ct + P_{R_1} - \Delta C_{R_1})$	$(PF - \Delta C_{cf_3} + P_{cf_3} - (cf_3 - f_2)Ct - (R_3 - r_1)Cr,$ $PF - (R_3 - r_1)Cr - (cf_3 - f_2)Ct + P_{R_2} - \Delta C_{R_2})$	$(PF - \Delta C_{cf_3} + P_{cf_3} - (cf_3 - f_2)Ct - (R_3 - r_2)Cr,$ $PF - (R_3 - r_2)Cr - (cf_3 - f_2)Ct + P_{R_3} - \Delta C_{R_3})$

If both carbon emissions and risk are at the tolerable level (see the payoff matrix in Table 3.9), and if the expected profits P_{R1} and P_{cf1} from the sale of 'environmentally sound' products fully cover the technological innovation cost ΔC_{R1}, ΔC_{cf1} arising from the risk and carbon emissions reduction, the following equations apply:

$$-\Delta C_{R1} + P_{R1} + r_1 Cr > 0 \tag{3.19}$$

$$-\Delta C_{cf1} + P_{cf1} + f_1 Ct > 0 \tag{3.20}$$

The dominant strategy is that the manufacturers will attempt to reduce both the carbon emissions and risk to the 'acceptable level', as the payoff at the acceptable level is higher than that at the tolerable level. Here, a dominant strategy is defined as the optimal strategy for one player, no matter what the other player chooses [20]. Otherwise, the manufacturers will be reluctant to improve and the situation will remain at the 'tolerable level'.

When the carbon emissions are at the tolerable level while the risk remains intolerable (see Table 3.10), and if the following two equations are both satisfied:

$$-\Delta C_{cf1} + P_{cf1} + f_1 Ct > 0 \tag{3.21}$$

$$(r_1 - r_2)Cr + (P_{R2} - P_{R3}) - (\Delta C_{R2} - \Delta C_{R3}) > 0 \tag{3.22}$$

Manufacturers will reduce both the carbon emissions and risk to the acceptable level. On the other hand, if the expected profit cannot cover the technological cost of reducing the carbon emissions of products, $(P_{cf1} - \Delta C_{cf1}) + f_1 Ct < 0$, manufacturers will maintain the carbon emissions at the tolerable level, but decrease the risk from the intolerable to the tolerable level, because the expected payoff of risk reduction at the tolerable level is higher than at the acceptable level.

Table 3.9 Payoffs of carbon emissions and risk at tolerable region

Carbon emission		Risk	
		Tolerable (r_2) ALARP (As low as reasonably practicable)	
		Maintain	Reduce to acceptable level
Tolerable (f_2)	Maintain	$(PF - R_2Cr - cf_2Ct,$ $PF - R_2Cr - cf_2Ct)$	$(PF - (R_2 - r_1)Cr - cf_2Ct,$ $PF - (R_2 - r_1)Cr - cf_2Ct - \Delta C_{R_1} + P_{R_1})$
	Reduce to acceptable level	$(PF - R_2Cr - (cf_2 - f_1)Ct - \Delta C_{cf_1} + P_{cf_1},$ $PF - R_2Cr - (cf_2 - f_1)Ct)$	$(PF - (R_2 - r_1)Cr - (cf_2 - f_1)C_t - \Delta C_{cf_1} + P_{cf_1},$ $PF - (R_2 - r_1)Cr - (cf_2 - f_1)C_t - \Delta C_{R_1} + P_{R_1})$

Table 3.10 Payoffs of carbon emissions at the tolerable region and risk at the intolerable region

Carbon emission		Risk	
		Intolerable	
		Reduce to acceptable level	Reduce to tolerable level
Tolerable (f_2)	Maintain	$(PF - cf_2Ct - (R_3 - r_1)Cr,$ $PF - (R_3 - r_1)Cr -$ $cf_2Ct + P_{R_2} - \Delta C_{R_2})$	$(PF - cf_2Ct - (R_3 - r_2)Cr,$ $PF - (R_3 - r_2)Cr -$ $cf_2Ct + P_{R_3} - \Delta C_{R_3})$
	Reduce to acceptable level	$(PF - (R_3 - r_1)Cr - (cf_2 -$ $f_1)Ct - \Delta C_{cf_1} + P_{cf_1},$ $PF - (R_3 - r_1)Cr - (cf_2 -$ $f_1)Ct - \Delta C_{R_2} + P_{R_2})$	$(PF - (R_3 - r_2)Cr - (cf_2 -$ $f_1)Ct - \Delta C_{cf_1} + P_{cf_1},$ $PF - (R_3 - r_2)Cr - (cf_2 -$ $f_1)Ct - \Delta C_{R_3} + P_{R_3})$

Similarly, if the carbon emissions are at the intolerable levelwhile risk remains tolerable (see Table 3.11), the dominant strategy for manufacturers is to reduce both the carbon emissions and risk to the acceptable level, as long as both of the following two expressions are satisfied:

$$(f_1 - f_2)Ct + (P_{cf2} - P_{cf3}) - (\Delta C_{cf2} - \Delta C_{cf3}) > 0 \tag{3.23}$$

$$-\Delta C_{R1} + P_{R1} + r_1 Cr > 0 \tag{3.24}$$

Otherwise, manufacturers will maintain the risk at the tolerable level, but decrease the carbon emissions from the intolerable to the tolerable level.

If both carbon emissions and risk are at the intolerable level (See the payoff matrix in Table 3.12), and if both Eqs. (3.25) and (3.26) are satisfied, reduction of both carbon emissions and risk to the acceptable level is the dominant strategy for manufacturers, because the economic benefits of taking acceptable action are greater

Table 3.11 Payoffs of carbon emissions at the intolerable region and risk at the tolerable region

Carbon emission		Risk	
		Tolerable (r_2) ALARP (as low as reasonably practicable)	
		Maintain	Reduce to acceptable level
Intolerable	Reduce to acceptable level	$(PF - \Delta C_{cf_2} + P_{cf_2} -$ $(cf_3 - f_1)Ct - R_2Cr,$ $PF - R_2Cr - (cf_3 - f_1)Ct)$	$(PF - \Delta C_{cf_2} + P_{cf_2} -$ $(cf_3 - f_1)Ct - (R_2 - r_1)Cr,$ $PF - (R_2 - r_1)Cr - (cf_3 -$ $f_1)Ct + P_{R_1} - \Delta C_{R_1})$
	Reduce to tolerable level	$(PF - \Delta C_{cf_3} + P_{cf_3} -$ $(cf_3 - f_2)Ct - R_2Cr,$ $PF - R_2Cr - (cf_3 - f_2)Ct)$	$(PF - \Delta C_{cf_3} + P_{cf_3} -$ $(cf_3 - f_2)Ct - (R_2 - r_1)Cr,$ $PF - (R_2 - r_1)Cr - (cf_3 -$ $f_2)Ct + P_{R_1} - \Delta C_{R_1})$

Table 3.12 Payoffs of carbon emissions and risk at the intolerable region

Carbon emission		Risk	
		Intolerable	
		Reduce to acceptable level	Reduce to tolerable level
Intolerable	Reduce to acceptable level	$(PF - \Delta C_{cf_2} + P_{cf_2} - (cf_3 - f_1)Ct - (R_3 - r_1)Cr,$ $PF - (R_3 - r_1)Cr - (cf_3 - f_1)Ct + P_{R_2} - \Delta C_{R_2})$	$(PF - \Delta C_{cf_2} + P_{cf_2} - (cf_3 - f_1)Ct - (R_3 - r_2)Cr,$ $PF - (R_3 - r_2)Cr - (cf_3 - f_1)Ct + P_{R_3} - \Delta C_{R_3})$
	Reduce to tolerable level	$(PF - \Delta C_{cf_3} + P_{cf_3} - (cf_3 - f_2)Ct - (R_3 - r_1Cr,$ $PF - (R_3 - r_1)Cr - (cf_3 - f_2)Ct + P_{R_2} - \Delta C_{R_2})$	$(PF - \Delta C_{cf_3} + P_{cf_3} - (cf_3 - f_2)Ct - (R_3 - r_2)Cr,$ $PF - (R_3 - r_2)Cr - (cf_3 - f_2)Ct + P_{R_3} - \Delta C_{R_3})$

than those of taking tolerable action.

$$(f_1 - f_2)Ct + (P_{cf2} - P_{cf3}) - (\Delta C_{cf2} - \Delta C_{cf3}) > 0 \tag{3.25}$$

$$(r_1 - r_2)Cr + (P_{R2} - P_{R3}) - (\Delta C_{R2} - \Delta C_{R3}) > 0 \tag{3.26}$$

According to the above analysis, whether manufacturers are willing to reduce the risk and carbon emissions voluntarily is mainly determined by the cost and sales profit. If the additional sales profit can cover the technological innovation cost, both risk and carbon emissions will be reduced. Otherwise, manufacturers will keep either the risk or carbon emissions at the original level. In the next section, the effects of government punitive policies and incentives will be discussed.

3.4.4 A Constrained Game Between Government and Manufacturer

In this game scenario, there are two players who separately represent the government and the manufacturers of potentially 'hazardous materials' within the supply chain. The overall objective of reducing risk and carbon emissions, while still ensuring the economic interest of the supply chain, remains the focus of the game. To simplify the mathematical model, manufacturers will choose the same strategy under the specific government policy, i.e. to produce at least within the tolerable region. Let E and U denote the economic benefits that government would receive in terms of revenue if manufacturers produce at the tolerable and intolerable levels, respectively. C_s indicates the cost of government supervision. The game between the government and the manufacturers, when the risk or carbon emissions are in the intolerable level, is shown in Table 3.13. Government makes the decisions between P and NP, which indicates Penalty and No-penalty strategy, respectively. Meanwhile, manufacturers make their own decisions on whether to reduce the risk and carbon emissions to the tolerable level (RT) or to maintain (M) the current carbon emissions and risk level.

This game model consists of a three-stage game (see Fig. 3.18). In the first stage, when there is no government penalty (NP), the cost of government supervision is higher than the penalties ($C_s > P$). Additionally, the technological cost ($\Delta C_3 = \Delta C_{R3} + \Delta_{cf3}$) is quite high for the manufacturers to maintain by reducing both risk and carbon footprint ($r_2Cr + f_2Ct + P_3 < \Delta C_3$). Thus, maintaining the production mode at the current level (M) is the dominant strategy for the manufacturers.

In the second stage, if the government decides to penalize (P) the environmentally unfriendly action of the manufacturers, and if the economic sanctions are increasing gradually above the supervision cost, as $C_s < P$, the manufacturers will prefer to pay the economic penalties to maintain the current production mode (M) rather than to reduce both the risk and carbon emissions level, because if the penalty is

Table 3.13 Strategic game between government and manufacturers in the intolerable region

Order of payoffs Government, manufacturers		Manufacturers	
		RT	*M*
Government	*P*	$(E - C_S, PF + (P_{cf_3} - \Delta C_{cf_3}) + (P_{R_3} - \Delta C_{R_3}) - (cf_3 - f_2)Ct - (R_3 - r_2)Cr)$	$(U - C_S + P, PF - R_3Cr - cf_3Ct - P)$
	NP	$(E, PF + (P_{cf_3} - \Delta C_{cf_3}) + P_{R_3} - \Delta C_{R_3}) - (cf_3 - f_2)Ct - (R_3 - r_2)Cr)$	$(U, PF - R_3Cr - cf_3Ct)$

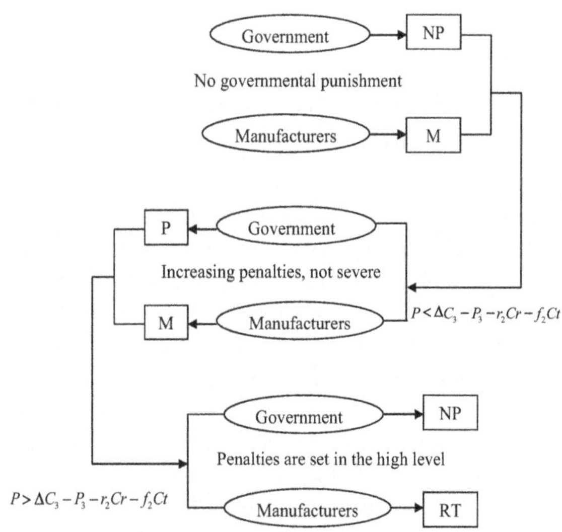

Fig. 3.18 Three stages in the punitive game between government and manufacturers

not severe, the inequality ($r_2Cr + f_2Ct + P_3 + P < \Delta C_3$) holds. Thus, the pair of dominant strategy has been transformed into 'P-M'.

Assuming, in the third stage, that the governmental penalties are set high enough, the manufacturers will choose to change their existing production mode (RT). The level of penalty should satisfy by the following equation:

$$P > (\Delta C_{R3} + \Delta C_{cf3}) - (P_{R3} + P_{cf3}) - r_2Cr - f_2Ct \tag{3.27}$$

Moreover, if the technological cost can be fully covered by the additional sales of 'environmentally friendly product', then $r_2Cr+f_2Ct+P_3-\Delta C_3 > 0$, and the government will no longer need to take action to regulate the process (P), the manufacturers will choose to reduce the risk and carbon emissions to the tolerable level (RT). Thus, the 'win–win' strategy can be 'NP-RT'.

3.4.5 An Incentive Game Between Government and Manufacturer

In order to promote the supply chain to be much 'cleaner' and 'greener', the government should also introduce some incentive policies to help the manufacturers upgrade their current production method [67]. Based on the premise that production is prohibited if either risk or carbon emissions are at the intolerable level, this study only discusses how to incentivize the manufacturers to improve the carbon emissions and risk from the tolerable to the acceptable level.

Let I denote the government incentive in the form of subsidies i.e. through tax breaks, and V indicate the additional government revenue obtained by the manufacturers producing at the acceptable level. The game between the government and the manufacturers is a two-stage game, Table 3.14, as shown in Fig. 3.19. Suppose that the government initially chooses the strategy I (Incentive) or NI (No Incentive). In response to these options, the manufacturers have to decide whether to reduce their carbon emissions and risk to the acceptable level (RA) or to maintain at the current level (M).

Table 3.14 Strategic game between government and manufacturers in the tolerable region

Order of payoffs Government, manufacturers		Manufacturers	
		RA	*M*
Government	*I*	$(V - I, PF + (P_{cf_1} - \Delta C_{cf_1}) + (P_{R_1} - \Delta C_{R_1}) - (cf_2 - f_1)Ct - (R_2 - r_1)Cr + I)$	$(-I, PF - R_2Cr - cf_2Ct)$
	NI	$(V, PF + (P_{cf_1} - \Delta C_{cf_1}) + (P_{R_1} - \Delta C_{R_1}) - (cf_2 - f_1)Ct - (R_2 - r_1)Cr)$	$(0, PF - R_2Cr - cf_2Ct)$

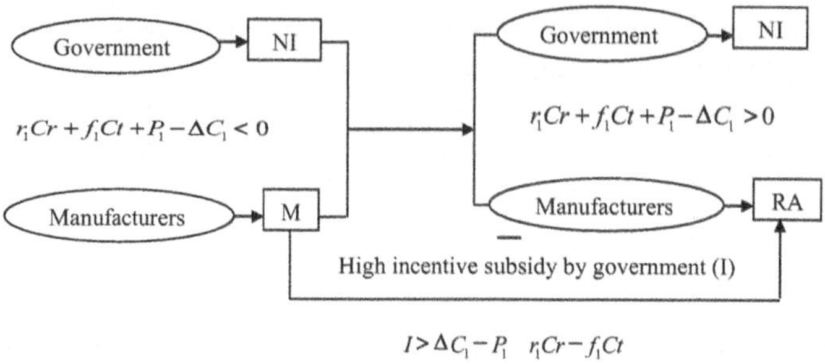

Fig. 3.19 Two stages in the incentive game between government and manufacturers

From Table 3.14, to give any subsidy to encourage manufacturers to reduce the risk and carbon emissions should be the dominated strategy for government. Thus, the government prefers to choose 'NI' rather than 'I'. In the first stage, if the government chooses the strategy 'NI', and if $r_1 Cr + f_1 Ct + (P_{R1} + P_{cf1}) - (\Delta C_{R1} + \Delta C_{cf1}) < 0$, where the technological innovation cost ($\Delta C_1 = \Delta C_{R1} + \Delta C_{cf1}$) cannot be fully covered by the additional sales profits ($P_1 = P_{R1} + P_{cf1}$) and the cost savings from risk and carbon emissions reduction to the acceptable level ($C_S = r_1 Cr + f_1 Ct$), manufacturers will choose to maintain production in the tolerable region. However, if the government chooses to incentivize ('I') the manufacturers to reduce, and the incentive subsidy satisfies the following expression:

$$I > (\Delta C_{R1} + \Delta_{cf1}) - (P_{R1} + P_{cf1}) - r_1 Cr - f_1 Ct \tag{3.28}$$

The manufacturers will tend to reduce risk and carbon emissions from the tolerable level to the acceptable level. Otherwise, if the incentive subsidy is not high enough, manufacturers will choose to maintain current practices.

The second stage suggests that, if $r_1 Cr + f_1 Ct + (P_{R1} + P_{cf1}) - (\Delta C_{R1} + \Delta C_{cf1}) > 0$, where the technological innovation cost can be fully recovered, the manufacturers are willing to decrease the risk and carbon emissions to the acceptable level, so that the government incentive is not necessary at this stage. Thus, the strategic action pair has been transformed into ('NI-RA'), which is now the 'win–win' strategy for both parties.

3.4.6 An Illustrative Case

In order to show how game theory informs the national manufacturers in the supply chain to reduce risk and carbon emissions, an illustrative example is presented in this section. Here, the data is mainly taken from the annual reports of a large company [51] and the corporate responsibility report [51]. To demonstrate the actions of manufacturers without government supervision, it is assumed that the risk is initially at the tolerable level within the current supply chain, whilst the carbon emissions remain intolerable. Table 3.15 shows the game model parameters chosen for the case example. This allows us to quantify the payoff matrix shown in Table 3.11.

References Wills and Finney (2004), Green (2006), UK HM Revenue and Customs (2008), Tesco (2010, 2011) and The Carbon Trust (2010) are cited in the text but not provided in the reference list. Please provide the respective references in the list or delete these citations. All the citations have been updated

Moreover, a game theory tool based on Visual Basic programming is developed to solve the two persons pure and mixed strategy game, where the quantified payoff matrix can be seen in Fig. 3.20. The solution suggests a dominant strategy in this game, which is to reduce the carbon emissions to the acceptable levelwhile keeping the risk at the tolerable level. The dominant strategy is strongly influenced by the purchase preference of customers who choose environmentally friendly products. If the customers prefer environmentally friendly products more than existing products, the demands should be increased to reduce the carbon emissions. For instance, the cited group has established through surveys that 59% of UK customers bought at least one carbon labeled product []. However, the HSE considers in reducing risk that it is reasonable to spend up to three times the value of the overall benefit of the risk reduction [26], which highlights the difficulty of reducing the environmental risk at the same time.

3.4.7 Summary

Game theory provides a valuable tool in the identification of the dominant strategy for improving performance along the three dimensions of the Triple Bottom Line (Economic, Society and Environment). This study has provided a novel perspective in applying game theory to the green supply chain for selecting appropriate strategies to limit the potential environmental impacts, and by inclusion of the principle of the 'tolerability of risk'. According to the analysis, different strategies have been initially identified and compared in order to reduce the risk, and thus the carbon emissions, while maximizing the economical benefits. Moreover, major factors have been identified by the construction of the game model, such as cost and benefit, government punitive and incentive policy etc., to provide insight into the problem situation.

Based on the game theoretical analysis, the potential actions of manufacturers within the different scenarios are given as follows. In the absence of government supervision, the dominant strategy of the manufacturers will mainly depend on the difference between the payoff of the re-designed supply chain and the payoff of reducing carbon emissions and risk to the specified level, e.g. tolerable level, acceptable level. If the payoff after reducing carbon footprint or risk throughout the supply chain is shown to be greater than the cost, , the strategy that manufacturers are expected to adopt is to implement improvements in production methods. Otherwise, the strategy is to maintain the original level of risk and carbon emissions.

Table 3.15 Model parameters of the case examples

Parameters	Numerical value	Source
PF	60,931 million GB Pounds	Investigated group revenue at 2010 (exc.VAT)
cf_3	5.44 million tonnes	The carbon emissions measured by the Group at 2011
f_1	0 tonne	Tesco has claimed to be a 'zero-carbon' business by 2050
f_2	3.81 million tonnes	As the carbon emissions of the products in the Group supply chain is claimed to be reduced by 30% at 2020, f_2 can be derived by the current carbon emissions (5.44 million tonnes) to multiply 70%
Cr	10^6 GB pounds	A value of preventing a fatality (VPF) related to a reduction in risk of one in a hundred thousand
Ct	£12/tonne CO_2	The Carbon Trust [76] suggests that the carbon price is fixed as£12/tonne
R_2	10^{-5}	Assume the risk is at the tolerable level
r_1	10^{-6}	The risk criterion between tolerable and acceptable region
ΔC_{cf3}	22.08 million GB pounds	The Group has launched a green credit program, worth over £10 million to incentivize green behaviours of the consumers each year, which subordinates to the group cost of sales. We have calculated the average growth rate of cost of sales, as 9.2%. Thus, the cost for carbon emissions reduction to year 2020 can be measured as $10 \times (1 + 9.2\%)^9 = 22.08$ million GB pounds, whilst selecting year 2011 as the baseline
ΔC_{cf2}	309.52 million GB pounds	We still use the cost of 2011 as the baseline, the cost for carbon emissions reduction to the year 2050 can be measured as $10 \times (1 + 9.2\%)^{39} = 309.52$ million GB pounds
P_{cf3}	185.36 million GB pounds	The Group has carbon labeled about 1100 products since 2008, which account for 1.3 billion of sales each year. As the group operating profit margin is 6.3% at 2011, we calculate the 'green' products profit is $1300 \times 6.3\% = 81.9$ million pounds in 2011. In addition, we have calculated the average growth rate of the sales profit as 9.5%. Thus, sales profit at year 2020 can be measured as $81.9 \times (1 + 9.5\%)^9 = 185.36$ million GB pounds, whilst selecting year 2011 as the baseline
P_{cf2}	2821.20 million GB pounds	Sales profit at year 2020 can be measured as $81.9 \times (1 + 9.5\%)^{39} = 2821.20$ million GB pounds

(continued)

Table 3.15 (continued)

Parameters	Numerical value	Source
ΔC_{RI}	6411.32 million GB pounds	Risk management is a key issue in the investigated Group. We suggest the annual administrative expense to be fully used for risk control. HSE [26] suggests the cost discount rate is suggested by 3.5% for risk reduction. Thus, the cost for risk reducing from the tolerable to the acceptable level can be measured as $1676 \times (1 + 3.5\%)^{39} = 6411.32$ million pounds, whilst selecting year 2011 as the baseline
P_{RI}	2137.11 million GB pounds	HSE suggests that it is reasonable to spend up to three times the value of the risk reduction, when applied to workers. Thus, the benefit for risk reducing from the tolerable to the acceptable level can be measured as $6411.32/3 = 2137.11$ million pounds

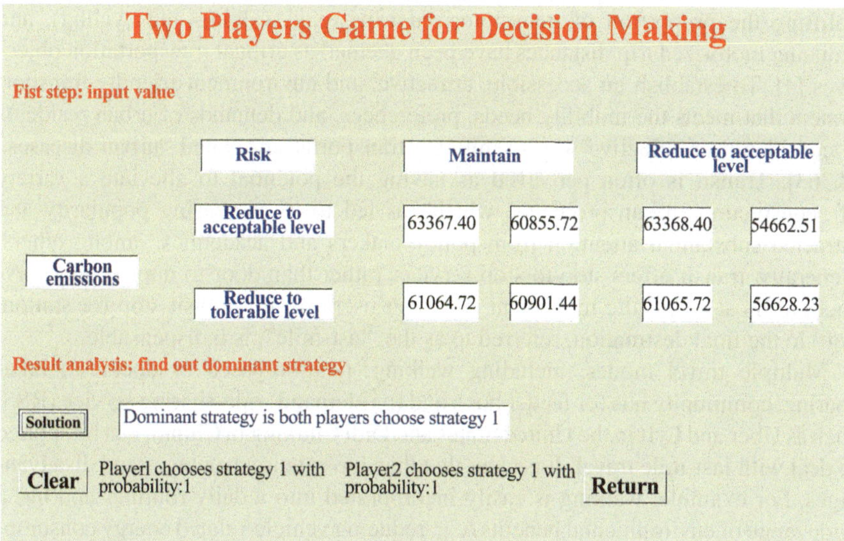

Fig. 3.20 Solution of case example in the designed game theory tool

The preferential purchase of environmentally friendly products by customers will greatly influence the dominant strategies of manufacturers when there is no supervision from the government [69]. However, if the government increases the penalties on the environmental unfriendly actions of manufacturers, the strategic choices of the manufacturers will be decided by the level of such penalties. If the penalty is slight and not large enough to induce the manufacturers to change the current risk and carbon emissions level, the dominant strategy is to maintain the current level. If the penalty is large enough, improving the risk and carbon emissions at least to the tolerable level should be the dominant strategy for manufacturers. Finally, if the government initiates a sanctions policy for the environmentally unfriendly actions of the manufacturers, the tendency will be to maintain the risk and carbon emissions level at best within the tolerable band at best. If the government aims to further improve the risk and carbon emissions to the broadly acceptable level, then an incentive policy will be required, such as a subsidy from government, which may promote the required risk and carbon emissions reduction [71].

3.5 A Last-Mile Travel Mode Selection

Cities around the world have been grappling with various conspicuous issues resulting from the excessive use of motor vehicles, including but not limited to traffic congestion, air/noise pollution, environmental degradation, limited green open space, and carbon emissions. In this context, decreasing the motorized trip frequency, uplifting the proportion of non-motorized trips (e.g. walking and cycling), and reducing motorized trip distances have been deemed as critical transportation objectives [4]. To establish an accessible, attractive, and environment-friendly transport system that meets the mobility needs, preferences, and demands of urban residents is considered an effective way to mitigate transportation-related "urban diseases" [4, 63]. Transit is often perceived as having the potential to alleviate a variety of conspicuous urban problems, which has led to its increasing popularity and attracted substantial attention from policy makers and academics, among others. Generally, transit offers stop-to-stop services rather than door-to-door services. As such, in most cases, the movement of transit users from a transit stop (or station, hub) to the final destination, referred to as the "last-mile", is indispensable.

Multiple travel modes, including walking, (station-based or dockless) bike-sharing, community bus (or feeder bus), and (on-demand) ride-sharing service (RSS, such as Uber and Lyft in the United States and Didi Chuxing in China), can be applied to deal with last-mile travel. It is clear that they have their advantages and disadvantages. For example, walking is easily incorporated into a daily routines and has a wide range of environmental benefits (e.g. reducing vehicle-related energy consumption), economic benefits (e.g. reducing fuel costs and travel costs), health benefits (e.g. lowering the risk of depression and obesity), and social benefits (e.g. increasing social capital, encouraging social interaction, and nurturing weak social ties) [13, 32, 35]. It, however, may only be applicable to short trips because of physical or mobility limitations. As Munoz-Raskin [41] proposed, pedestrians can only bear walking for a maximum of 10 minutes to access transit. In a similar vein, dockless bike-sharing, which is popular in many contexts (e.g. China and Singapore), has numerous advantages, such as flexibility and ease of pick-up and drop-off [65]. However, riders may park it arbitrarily, and such disorderly parking behavior may cause traffic congestion and other congestion related issues, such as lost productive time [65]. Likewise, the RSS is featured as a personalized, door-to-door and demand-responsive service, wide spatial/temporal coverage, flexibility, and time savings. Its fare, however, is fairly high, laying out an issue related to financial affordability.

The selection of travel modes can be considered a multiple attribute decision making (MADM) problem due to the trade-offs among various conflicting criteria. Its associated weighting methods include but are not limited to simple additive weighting method, weighted product method, Delphi method, analytic hierarchy process (AHP) method, multiplicative AHP method, and technique for order preference by similarity to ideal solution (TOPSIS) method. However, they have their own characteristics, advantages, and disadvantages. In other words, all of them are far from perfect and still face criticisms and challenges. For example, the AHP method is an easy-to-use but time-consuming criterion weighting method that allows hierarchical modeling of the problem. It was originally developed by Saaty in the 1970s [47]. Since then, the AHP method has received flourishing scientific attention and also some criticism (e.g. rank inversion) and thus has extensively been refined by numerous researchers [15]. However, it is highly dependent on the knowledge or experience of the resorted individual experts, though it has been substantially improved.

This section proposes a new approach by a hybrid of data mining with multiple attribute decision making to identify the optimal travel mode for the last mile, where data mining is applied to objectively determine the weights. Methodologically, this study clearly shows that the combination of the two techniques (i.e. data mining and MADM) creates new possibilities for investigating many important, existing or emerging issues. Four last-mile travel modes, including walking, bike-sharing, community bus, and on-demand ride-sharing service, are ranked based on three evaluation criteria, namely, travel time, monetary cost, and environmental performance. More specifically, four last-mile modes, including walking, bike-sharing, community bus (abbreviated as "bus" hereinafter), and RSS, serve as decision making units and constitute the choice set; three indicators, namely, travel time, monetary cost, and environmental performance are set as evaluation criteria; and Sina Weibo data are used to determine the weights of the proposed criteria (data mining).

The study area is Chengdu, a sub-provincial, inland city located in the southwest part of China and the western part of the Sichuan Basin. It is the capital and the educational, financial, cultural, business, trade, technology, and transportation center of Sichuan Province and an economic hub of western China. The findings are expected to improve the efficiency of the transit services and provide insights into policy making in last-mile transportation planning and management.

3.5.1 Construction of an Indicator System

Travel time and travel (monetary) cost have been identified as the most important factors in travel mode choice or selection in a large body of existing literature [49], predominantly because the two service level variables serve as the strongest and the most negatively perceived disincentives. Traditionally, the identification of the optimal travel mode is mainly based on travel time and/or travel cost [49]. This, however, is inadequate in the era of the increasing awareness of climate change and carbon emissions. In other words, carbon emissions reduction should be regarded

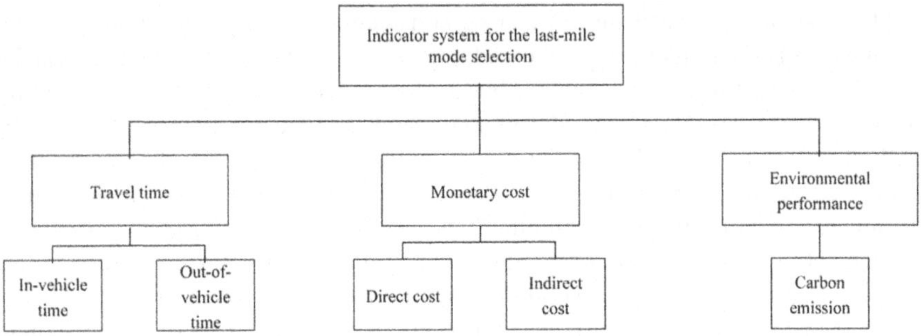

Fig. 3.21 Indicator system for last-mile travel mode selection

as an additional criterion for determining the optimal travel mode. Existing literature provides strong evidence to support such an implication that the travel mode shift has the potential to contribute to carbon emissions reduction. In such context, the last-mile travel modes in this study mainly considered for the selection include walking, bike-sharing, bus, and RSS, and three sets of evaluation criteria, namely, travel time, monetary cost, and environmental performance, are taken into evaluation. The indicator system for last-mile travel mode selection is shown in Fig. 3.21.

Travel Time

The travel time of walking for last-mile (T_w) is measured as follows

$$T_w = \frac{s}{v_w} \tag{3.29}$$

where s is the travel distance (km) and v_w is the average walking speed (km/min).

Different from walking, the travel time of bike-sharing, the bus, and the RSS includes in-vehicle time and out-of-vehicle time. The travel time of bike-sharing for the last-mile (T_b) is measured as follows:

$$T_b = t_{bo} + \frac{s}{v_b} \tag{3.30}$$

where t_{bo} is the bike-searching time (or out-of-bike time) (min); v_b is the average cycling speed (km/min); and s is as defined before.

The travel time of the bus for last-mile (T_c) is measured as follows:

$$T_c = t_{co} + \frac{s}{v_c} \tag{3.31}$$

where t_{co} is the wait time (or out-of-vehicle time) (min); v_c is the average bus travel speed (km/min); and s is as defined before.

The travel time of the RSS for last-mile (T_r) is measured as follows:

$$T_r = t_{ro} + \frac{s}{v_r} \tag{3.32}$$

where t_{ro} is the wait time (or out-of-vehicle time) (min); v_r is the average RSS travel speed (km/min); and s is as defined before.

Monetary Cost

The monetary cost includes direct cost and indirect cost. The former indicates the per-trip cost, such as the travel fare, while the latter refers to the one-off cost, such as the cost of bus card purchase and a bike-share deposit.

The monetary cost of walking is set as 0.

The monetary cost of bike-sharing (M_b) is expressed as follows:

$$M_b = \begin{cases} m_b if \frac{s}{v_b} \leq t \\ m_b + m_c if \frac{s}{v_b} > t \end{cases} \tag{3.33}$$

where m_b is the initial charge (or basic fee); t is the maximum use time (e.g. 1 h) with the initial charge (e.g. 1 Yuan); m_c is the additional charge (Yuan) if the riding time exceeds t; and s is as defined before.

The monetary cost of the bus (M_c), which assumes to adopt a flat (or uniform) fare structure in this study, is expressed as follows:

$$M_c = \frac{a}{r_z} + b \tag{3.34}$$

where a is the bus card capital cost (or purchasing cost) (Yuan); r_z is the average use time of a bus card throughout its life; and b is the flat (or constant) travel price (Yuan).

The RSS normally adopts a non-linear fare structure, which in this study is assumed to be correlated with both travel distance and travel time. If the travel distance and travel time are less than the predetermined thresholds, only the initial fare (e.g. 9 Yuan in Chengdu) needs to be paid. Otherwise, additional charges must be paid. As such, the monetary cost of the RSS (M_t) is expressed as follows:

$$M_r = max(U, s \times m_d + \frac{s}{v_r} \times m_t) \tag{3.35}$$

where m_d is the cost per kilometer (Yuan/km); m_t is the cost per min (Yuan/min); U is the initial charge (Yuan); and other variables are as defined before.

Environmental Performance

The environmental performance of last-mile travel is measured in terms of direct carbon emissions. Walking and bike-sharing are not directly related to the burning of fossil fuels, so their direct carbon emissions are set to be 0 in this study.

The carbon emissions of bus travel (E_c) can be expressed as follows:

$$E_c = \frac{c_c \times e_c \times s}{q} \tag{3.36}$$

where c_c is the carbon emissions factor (kg/GJ) for the bus fuel; e_c is the average fuel consumption (GJ) per kilometer of bus travel; q is the average passenger volume of the bus; and s is as defined before;

The carbon emissions of RSS travel (E_r) can be expressed as follows:

$$E_r = \frac{c_r \times e_r \times s}{p} \tag{3.37}$$

where c_r is the carbon emissions factor (kg/GJ) for the RSS fuel; e_r is the average fuel consumption (GJ) per kilometer of RSS travel; p is the average passenger volume of the RSS; and s is as defined before.

3.5.2 A Data Mining Driven Indicators Weighting

In this section, a data mining-based weighting method and MADM are used to rank the four last-mile modes (Fig. 3.21). Generally, weighting methods can be roughly categorized into two groups, namely, subjective method (e.g. AHP method and multiplicative AHP method) and objective method (e.g. simple additive weighting method, weighted product method, entropy method, neural network method, standard deviation method, data envelopment analysis, and TOPSIS method) [23]. Both categories of methods, however, have their own specific shortcomings. In this section, the weights of all criteria were obtained by the keyword frequency, which likely guarantees their objectivity.

Data mining used the "Weibo search" function of Sina Weibo, and the four travel modes were set as keywords. Sina Weibo is a popular online social networking platform in China (like Twitter in the United States), through which users can publish microblogs (*weibos*) and disseminate, share, and receive information. It accounts for a high proportion of Weibo users in China, and its users cover a broad spectrum of Chinese people. This feature provides detailed sample data support for this section.

The data used for the subsequent analysis covered the period from January 1, 2017 to December 31, 2017. Using the Octopus network data collector, data crawling rules were written in the custom mode, and Weibo data were automatically obtained and saved. Finally, 54,429 Weibo observations were obtained by using specified keywords for search, and the BlueMC platform was used to analyze the word frequency. The keywords with the first 1000 frequencies were selected, and the keywords related to each indicator were retrieved. The word frequency was used as the index weight, which can be expressed as $W = [w_1, w_2, w_3]$.

Since the selected criteria may have different dimensions, it is necessary to normalize the data to a common scale for comparison. The attribute matrix (A) for each travel distance (1, 2, 3, 4, 5 km in this study) is constructed as follows:

$$\mathrm{A} = (a_{ij})_{n\times 3} \tag{3.38}$$

where i represents the travel mode index, and $i = 1, 2, 3, 4, 5, \ldots$, n represents walking, bike-sharing, the bus, the RSS, and various possible travel mode combinations (e.g. walking + bus), respectively;$j = 1, 2, 3$ indicates travel time, monetary cost, environmental performance of travel mode i, respectively; and a_{ij} is the value of criteria j of travel mode i [see Eqs. (3.29)–(3.37)].

The transformation rule used to convert indicator a_{ij} to standardized indicator $\tilde{a_{ij}}$ was as follows:

$$\tilde{a_{ij}} = \frac{a_{ij} - \mu_j}{s_j} \tag{3.39}$$

where μ_j, s_j represent the mean and the standard deviation of criteria j, respectively. Mathematically, μ_j, s_j can be expressed as follows:

$$\mu_j = \frac{1}{n}\sum_{i=1}^{n} a_{ij} \tag{3.40}$$

$$s_j = \sqrt{\frac{1}{n-1}\sum_{i=1}^{n} (a_{ij} - \mu_j)^2} \tag{3.41}$$

After standardization, the normalized attribute matrix becomes $(\tilde{a_{ij}})_{n\times 3}$.

As of the end of 2017, Chengdu has a total population of 14.35 million, and it covers an area of approximately 14.34 thousand km^2, consisting of eleven districts, five county-level cities, and four counties. In recent years, the Chengdu government has taken a series of transportation policies aimed at effectively meeting diversified travel demands and improving the quality of last-mile travel services. In this Context, the role of last-mile travel services in the transportation system has been greatly emphasized and strengthened, and the services have been gradually improved.

By investigating the last-mile travel modes of Chengdu residents, this study comprehensively evaluates and ranks four travel modes and various travel mode combinations in different travel distances to verify the effectiveness of the proposed data mining based weighting and MADM coupling method.

The weights of the three criteria were derived by counting the corresponding keywords. Table 3.16 presents the results of the data mining based weighting. Keywords in the category of travel time received the most attention from Sina Weibo users (frequency = 11,757), followed by those in the monetary cost (frequency = 7046). The weight of travel time, monetary cost, and environmental performance is 0.49, 0.29, and 0.22, respectively.

The parameter values in Eqs. (3.43)–(3.51) were obtained from field investigation, government policy documents, research reports, etc. Their details are shown in Table 3.17.

Table 3.16 Results of data-mining-based weighting

Travel time		Monetary cost		Environmental performance	
Keyword	Frequency	Keyword	Frequency	Keyword	Frequency
Convenience (*fangbian, bianli, bianmin*, or *bianjie*)	8175	Deposit (yajin)	3,228	Green (*lvse*)	374
Rapid (*kuaisu, gaosu*, or *kuai*)	2294	Economy (*jingji*)	1,688	Alternative energy (*xin nengyuan*)	584
Speed (*sudu*)	315	Free (*mianfei*)	907	Energy-saving (*jieneng*)	347
Slow (*man*)	228	Hire (*zulin*)	593	Pollution (*wuran*)	289
Inconvenience (*bubian*)	477	Charge (*shoufei*)	352		
Efficiency (*xiaolv*)	268	1 Yuan (*yiyuan*)	278		
Sum	11,757	7046		5194	
Proportion	0.49	0.29		0.22	

3.5.3 Results

Five scenarios with a last mile travel distance of 1, 2, 3, 4, or 5 km were considered, and four travel modes and a variety of their combinations were ranked according to the proposed method. Figure 3.22 shows the results of two scenarios (travel distance = 1 or 2 km). In both scenarios, bike-sharing is ranked first. On the one hand, it has a significant speed advantage over walking, 2.6 times faster than walking. On the other hand, the out-of-vehicle time is substantially shorter than that for the bus and the RSS. Moreover, it has an unrivaled advantage in the area of carbon emissions compared to the bus and the RSS during the use process.

Figure 3.23 reveals the results of the other three scenarios (travel distance = 3, 4, or 5 km). As last-mile distance is relatively long, multiple modes (e.g. walking + bus) for a trip are considered. Figure 3.23a shows that in the 3-km trip scenario, the best mode choice is still bike-sharing, closely followed by the bus. The use of multiple modes can alleviate the disadvantages of single travel modes and further improve travel efficiency in some cases. It, however, may involve two out-of-vehicle times (e.g. one time for bike-searching and one time for bus waiting), so it does not have a high ranking.

Figure 3.23b and c show that the community bus becomes the best travel choice for the 4-km and 5-km last-mile trips. This finding agrees with expectations and is consistent with the observation in other contexts, which indicates that people prefer to use the motorized mode (rather than walking and cycling) for long-distance last-mile trips [40]. Travel time differences (mainly due to speed differences) between modes

Table 3.17 Parameter values used in this study

Travel mode	Parameter	Value	Source
Walking	v_w	0.073 km/min	*Road Capacity Manual (daolu tongxing nengli shouce)*
Bike-sharing	t_{bo}	2.38 min	Field experiment
	v_b	0.1895 km/min	Field experiment
	t	$\geq$ 30 min	Field investigation
	m_b	1 Yuan	Field investigation
Bus	t_{co}	4.97 min	Field investigation
	v_c	0.333 km/min	*Implementation Opinions of the Priority Development of Urban Transit in the Chengdu Government (chengdushi renmin zhengfu guanyu youxian fazhan chengshi gonggong jiaotong de shishi yijian)*
	a	20 Yuan	Field investigation
	r_z	$\gg$ 20	Field investigation
	b	0 Yuan	Field investigation
	c_c	15.3 kg/GJ	*Provincial Guidelines for Greenhouse Gas Inventories (shengji wenshi qiti qingdan bianzhi zhinan)*
	e_c	10.91 MJ/km	*Patterns and Load of Gas Emissions by the CNG Automobile in Chengdu (chengdushi CNG qiche yongqi guilv ji yongqi fuhe)*
	q	22.95	Calculated based on *Chengdu Transportation Development Annual Report 2016*
RSS	t_{ro}	5.6 min	*Didi Smart Travel Big Data Report*
	m_d	1.6 Yuan/km	Field investigation
	v_r	0.593 km/min	*Didi 2017 Urban Traffic Travel Report*
	m_t	0.3 Yuan/min	Field investigation
	U	9 Yuan	Field investigation
	c_r	18.9 kg/GJ	*Guidelines for Provincial Greenhouse Gas Inventories*
	e_r	0.180 GJ/km	Field investigation
	p	2 passengers	Field investigation

become more pronounced as the travel distance increases. With the increase of travel distance, the speed disadvantage of bike-sharing (compared to that of the bus and the RSS) gradually emerges. Although the speed of the RSS is higher than that of the bus, the monetary cost is higher. In addition, the bus can carry more passengers, and the direct carbon emissions per passenger are lower than the RSS.

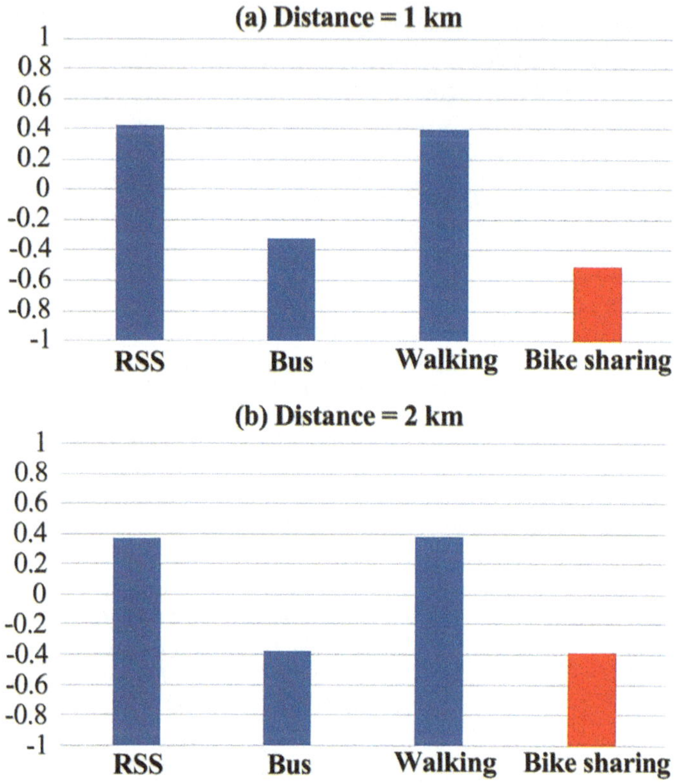

Fig. 3.22 Mode ranking for 1- and 2-km trips

Although walking involves no out-of-vehicle time and monetary cost and produces no emissions, it has obvious disadvantages in terms of travel speed and travel time. This becomes more apparent as the distance increases, as shown in Figs. 3.22 and 3.23.

According to the results of this study, it is suggested that governments, urban transport planners, and operators improve bike-sharing (for short last-mile trips) and the community bus service (for long last-mile trips). A few years ago, only station-based bike-sharing was available, requiring people to use a smart card to unlock, use, and pay for it, and the associated registration process was overly complicated. But dockless bike-sharing has gradually replaced station-based bike-sharing in numerous Chinese cities [28, 29, 33, 36]. The introduction of smartphone payments has become the norm in contemporary China, substantially simplifying the previously complicated registration process. Dockless bike-sharing has the potential to effectively meet at least a large proportion of the demand for short-distance last-mile travel.

Bike-sharing operators can propose measures or strategies to encourage the use of bike-sharing for last-mile trips, especially for short distance (< 3 km). As an

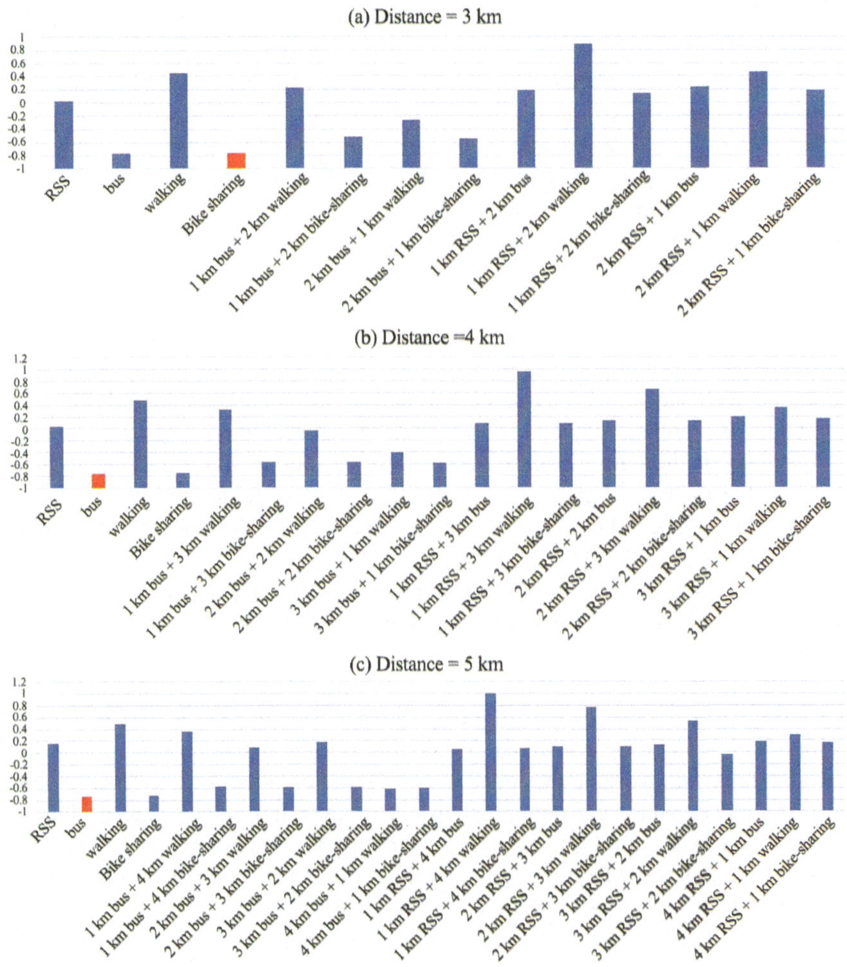

Fig. 3.23 Mode ranking for 3-, 4-, and 5-km trips

example, the "Call a Bike" program developed by German Railway in Berlin aims to encourage transit riders to use bike-sharing for first- or last-mile trips. The system intends to send an access code to bike-sharing users upon their request via phone calls or smartphone applications [44]. In addition, discounts are available for bike-sharing users if they combine the membership with the annual transit pass [1]. As a bonus for using bike-sharing for the last-mile, transit users can be rewarded with additional free service time for using bike-sharing , such as 90 min in Hangzhou, China. However, such strategies involve at least two operators (transit and bike-sharing), so the inter-operator coordination and integration in terms of ticketing or revenue sharing are indispensable. Furthermore, to decrease the number of idle bikes and increase the fluidity of bike-sharing, frequent rebalancing or adjusting bikes according to real-time demand, particularly during peak hours, is necessary [16].

The findings also implicate that it is important to propose planning strategies to encourage the use of bike sharing in the last-mile . Griffin and Sener [22] proposed a framework for collaborative and intermodal planning to strengthen the connection between bike-sharing and transit . They addressed that planners can leverage information on existing transport systems for planning at the local and regional level when expanding a bike-sharing system. For instance, in Chicago and Austin, some additional improvements to the bicycle network are needed, such as adding protected bike lanes to existing roadways. However, planning strategies for the integration of dockless bike-sharing–transit can be different from the traditional station-based bike-sharing–transit integration. Given that dockless bike-sharing modes are densely distributed near transit (e.g. metro) stations and are often parked randomly [65], the main focus of planning may be the regulation of disordered parking behaviors and improvement of the parking space near transit stations to achieve efficient dockless bike-sharing–transit integration. Related strategies include the installation of electronic fencing and the recommendation of parking areas (marked with a white rectangle on the ground) [29].

The community bus for long-distance last-mile trips (equal to or over 4 km) is supported by the findings. The community bus is a type of paratransit that operates on the borderline between public and private transportation in terms of cost and quality of service. It complements and serves areas without conventional public transit, primarily as a service gap filler [5, 45]. However, due to its nature, paratransit has long been ignored by traditional, formal transportation planning and policy. It is proposed that its network be expanded to serve more communities, especially those that are under served by conventional transit. Moreover, it is demand-responsive and may not have fixed routes and schedules, allowing for time-varying services. For example, operators can flexibly adjust the bus routes and schedules to effectively meet the mobility needs of last-mile travelers.

In order to guide residents to choose low-carbon and environmentally friendly travel modes, it is proposed to introduce a "Carbon Credits" system to quantify the use of low-carbon travel modes in the form of credits [68, 70]. For example, if residents choose to travel by bike-sharing or the bus, they can earn some credits. When the credits are accumulated to a certain amount, rewards (such as discounts for bike-sharing use) can be offered to encourage carbon emissions reduction. Such an approach is of paramount importance, especially in an era of increasing environmental awareness around the world [2].

It should be noticed that the discussions presented above are based on today's transport system, as the transport modes available to urban residents are still limited in Chengdu, China. Introducing "new" transport options is, therefore, both indispensable and urgent. Luckily, a large number of novel matters that have recently come into the transport sector (e.g. connected and autonomous vehicles, or smart connected vehicles and ubiquitous artificial intelligence applications) have the potential to further satisfy last-mile travel demands and drastically change many aspects of the transport system [64]. As such, they should arouse great fascination among local and international transportation researchers.

3.5.4 Summary

In this Section, a data mining based MADM for the last-mile travel mode selection is proposed. Four last-mile travel modes, including walking, bike-sharing, community bus, and RSS in Chengdu, China are ranked using three sets of criteria (i.e. travel time, travel cost, and low carbon performance). The results show that the optimal travel mode varies greatly with the distance of the "last-mile"; and that bike-sharing serves as the optimal travel mode if the last-mile distance is not more than 3 km, whilst the community bus becomes the optimal mode if the distance is between 4 and 5 km.

In general, transit offers stop-to-stop services rather than door-to-door service, and its takers must travel to their final destination after departing from a transit hub. As such, last-mile travel becomes an inevitable, essential component of transit travel. Its mode choice attracts the attention of policy makers and academics. Identifying the optimal mode for last-mile travel (e.g. walking and bike-sharing) can be considered as a MADM problem, which involves the determination of weights according to some methods. However, the methods, of either subjective or objective in nature, have their own specific characteristics, strengths, and drawbacks. It is expected that this study offers an evidence-based approach to guide residents in choosing the reasonable last-mile travel mode and provides insight into the development of a sustainable urban transport system.

References

1. Bachand-Marleau J, Lee BH, El-Geneidy AM (2012) Better understanding of factors influencing likelihood of using shared bicycle systems and frequency of use. Transp Res Record J Transp Res Board 2314:66–71
2. Balsa-Barreiro J, Li Y, Morales A (2019) Globalization and the shifting centers of gravity of world's human dynamics: implications for sustainability. J Clean Prod 239:117923
3. Bouder F, Slavin D, Löfstedt R (2007) The tolerability of risk: a new framework for risk management. Earthscan, London, pp 87–105
4. Cervero R, Kockelman K (1997) Travel demand and the 3Ds: density, diversity, and design. Transp Res Part D Transp Environ 2:199–219
5. Cervero R, Golub A (2007) Informal transport: a global perspective. Transp Policy 14:445–457
6. Chew I, Tan RR, Foo D, Chiu A (2009) Game theory approach to the analysis of inter-plant water integration in an eco-industrial park. J Clean Prod 17:1611–1619
7. China NPC (National People's Congress) (2007) Corporate income tax law of Peoples' Republic of China. China Legal Publishing House, Beijing (in Chinese)
8. China NPC (National People's Congress) (2012) Cleaner production promotion law of Peoples' Republic of China. China Legal Publishing House, Beijing (in Chinese)

9. Clay HH, Bassett WH (1999) Clay's handbook of environmental health. E&FN Spon, London, pp 451–461
10. DEFRA (Department for Environment Food and Rural Affairs). (2009) Guidance on how to measure and report your greenhouse gas emissions. http://www.defra.gov.uk/environment/business/reporting/pdf/ghg-guidance.pdf
11. Dinner IM, Heerde Van HJ, Neslin SA (2014) Driving online and offline sales: the cross-channel effects of traditional, online display, and paid search advertising. J Mark Res 51(5):527–545
12. Dong XQ, Li CL, Li J, Wang J, Huang WT (2010) A game-theoretic analysis of implementation of cleaner production policies in the Chinese electroplating industry. Resour Conserv Recycl 54:1442–1448
13. Du Toit L, Cerin E, Leslie E, Owen N (2007) Does walking in the neighbourhood enhance local sociability? Urban Studies 44:1677–1695
14. Elkington J (1997) Cannibals with forks: the triple bottom line of 21 century business. Capstone, Oxford
15. Emrouznejad A, Marra M (2017) The state of the art development of AHP (1979–2017): a literature review with a social network analysis. Int J Prod Res 55:6653–6675
16. Faghih-Imani A, Eluru N (2016) Determining the role of bicycle sharing system infrastructure installation decision on usage: case study of montreal BIXI system. Transp Res Part A Policy Pract 94:685–698
17. Fuller CW, Vassie L (2004) Health and safety management: principles and best practice. Prentice Hall, Essex, pp 182–190
18. Fundenberg D, Tirole J (1991) Game theory. MIT Press, Cambridge, p 11
19. Gale D (2000) Strategic foundations of general equilibrium: dynamic matching and bargaining games. Cambridge University Press, Cambridge, p 219
20. Geckil IK, Anderson PL (2010) Applied game theory and strategic behaviour. Chapman & Hall, London
21. Gibbons R (1992) A primer in game theory. Prentice Hall, London, pp 8–14
22. Griffin GP, Sener IN (2016) Planning for bike share connectivity to rail transit. J Public Transp 19:1–22
23. Ho W, Xu X, Dey PK (2010) Multi-criteria decision making approaches for supplier evaluation and selection: a literature review. Eur J Oper Res 202:16–24
24. Hou AZ (2011) A case study on cleaner production for rare earth industry: taking Leshan Shenghe rare earth company as an enterprise. Master of thesis, Southwest Jiaotong University, Chengdu, China (In Chinese with English Abstract)
25. HSE (Health & Safety Executive) (1988) The tolerability of risk from nuclear power stations. http://www.hse.gov.uk/nuclear/tolerability.pdf
26. HSE (Health & Safety Executive) (2001) Reducing risks, protecting people: HSE'S decision making process. http://www.hse.gov.uk/risk/theory/r2p2.pdf
27. Ji P, Ma X, Li G (2015) Developing green purchasing relationships for the manufacturing industry: an evolutionary game theory perspective. Int J Prod Econ 166:155–162
28. Ji Y, Ma X, Yang M, Jin Y, Gao L (2018) Exploring spatially varying influences on metro-bikeshare transfer: a geographically weighted Poisson regression approach. Sustainability 10:1526
29. Jia L, Liu X, Liu Y (2018) Impact of different stakeholders of bike-sharing industry on users' intention of civilized use of bike-sharing. Sustainability 10:1437
30. Jiang H, Zhao S, Yuan Y, Zhang L, Duan L, Zhang W (2018) The coupling relationship between standard development and technology advancement: a game theoretical perspective. Technol Forecast Soc Chang 135:169–177
31. Kemp R, Parto S, Gibson RB (2005) Governance for sustainable development: moving from theory to practice. Int J Sustain Dev 8(1–2):12–30
32. Leyden KM (2003) Social capital and the built environment: the importance of walkable neighborhoods. Am J Public Health 93:1546–1551
33. Lin JJ, Zhao P, Takada K, Li S, Yai T, Chen CH (2018) Built environment and public bike usage for metro access: a comparison of neighborhoods in Beijing, Taipei, and Tokyo. Transp Res Part D Transp Environ 63:209–221

34. Liu C, Huang W, Yang C (2017) The evolutionary dynamics of china's electric vehicle industry—taxes vs. subsidies. Comput Indus Eng 113:103–122
35. Lund H (2002) Pedestrian environments and sense of community. J Plan Educ Res 21(3):301–312
36. Ma X, Ji Y, Jin Y, Wang J, He M (2018) Modeling the factors influencing the activity spaces of bikeshare around metro stations: a spatial regression model. Sustainability 10:3949
37. McKelyey RD, Palfrey TR (1995) Quantal response equilibria for normal form games. Games Econom Behav 10:6–38
38. McKelvey RD, McLennan AM, Turocy TL (2007) Gambit: software tools for game theory. http://www.gambit-project.org
39. McKelvey RD, McLennan AM, Turocy TL (2010) Gambit: software tools for game theory. http://www.gambit-project.org/doc/gui.html#investigating-dominated-strategies-and-actions
40. Meng M, Koh PP, Wong YD (2016) Influence of socio-demography and operating streetscape on last-mile mode choice. J Public Transp 19:38–54
41. Munoz-Raskin R (2010) Walking accessibility to bus rapid transit: does it affect property values? The case of Bogotá, Colombia. Transp Policy 17:72–84
42. Okabe M, Ohtani H (2009) Risk estimation for industrial safety in raw materials manufacturing. J Loss Prev Process Ind 22:176–181
43. OPSI (Office of Public Sector Information) (2008) Climate Change Act 2008. Chapter 27: Carbon target and budgeting. http://www.opsi.gov.uk/acts/acts2008/ukpga_20080027_en_2#pt1-pb1-l1g1
44. Paul F, Bogenberger K (2014) Evaluation-method for a station based urban-pedelec sharing system. Transp Res Part A Policy Pract 4:482–493
45. Rimmer P (1980) Paratransit: a commentary. Environ Plan A Econ Space 12:937–944
46. Romp G (1997) Game theory introduction and application. Oxford University Press, Oxford, pp 19–22
47. Saaty TL (1998) What is the analytic hierarchy process? In: Mathematical models for decision support. Springer, pp 109–121
48. Safarzyńska K, van den Bergh JC (2018) A higher rebound effect under bounded rationality: interactions between car mobility and electricity generation. Energy Econ 74:179–196
49. Salonen M, Broberg A, Kyttä M, Toivonen T (2014) Do suburban residents prefer the fastest or low-carbon travel modes? Combining public participation GIS and multimodal travel time analysis for daily mobility research. Appl Geogr 53:438–448
50. Skea J, Nishioka S (2008) Policies and practices for a low-carbon society. Climate Policy 8:S5–S16
51. Tesco Group. Tesco Annual Report (2007–2011). http://www.tescoplc.com/information-hub/document-centre/
52. Thomas LC (2003) Games, theory, and applications. Dover Publications, Mineola
53. Tian YH, Govindan K, Zhu QH (2014) A system dynamics model based on evolutionary game theory for green supply chain management diffusion among Chinese manufacturers. J Clean Prod 80:96–105
54. Tseng ML, Chiu ASF, Tan RR, Siriban-Manalang AB (2013) Sustainable consumption and production for Asia: sustainability through green design and practice. J Clean Prod 40:1–5
55. United Kingdom Parliament (2008) Climate Change Act 2008. http://www.legislation.gov.uk/ukpga/2008/27/pdfs/ukpga_20080027_en.pdf
56. United Nations (2009) Framework convention on climate change. http://unfccc.int/resource/docs/2009/sbi/eng/12.pdf
57. Van Zeijl-Rozema A, Cörvers R, Kemp R, Martens P (2008) Governance for sustainable development: a framework. Sustain Dev 16(6):410–421
58. Wang J (2010) A game theoretical analysis of implementation of cleaner production policies in electroplating industry. Master of thesis, University of Harbin Institute of Technology, Harbin, China (in Chinese with English Abstract)
59. Wang HF, Gupta SM (2011) Green supply chain management: product life cycle approach. McGraw-Hill, New York

60. Wang Y (2007) Combining data mining and game theory in manufacturing strategy analysis. J Intell Manuf 18:505–511
61. Wang Y, Huscroft JR, Hazen BT, Zhang M (2018) Green information, green certification and consumer perceptions of remanufactured automobile parts. Resour Conserv Recycl 128:187–196
62. Xu L, Su J (2016) From government to market and from producer to consumer: transition of policy mix towards clean mobility in China. Energy Policy 96:328–340
63. Yang L, Zhou J, Shyr OF, Huo DD (2019) Does bus accessibility affect property prices? Cities 84:56–65
64. Yap MD, Correia G, Van Arem B (2016) Preferences of travellers for using automated vehicles as last mile public transport of multimodal train trips. Transp Res Part A Policy Pract 94:1–16
65. Zhao D, Wang D (2019) The research of tripartite collaborative governance on disorderly parking of shared bicycles based on the theory of planned behavior and motivation theories—a case of Beijing, China. Sustainability 11:5431
66. Zhao R, Deutz P, Neighbour G, McGuire M (2012) Carbon emissions intensity ratio: an indicator for an improved carbon labelling scheme. Environ Res Lett 7:014014
67. Zhao R, Zhou X, Han JJ, Liu CL (2016) For the sustainable performance of the carbon reduction labeling policies under an evolutionary game simulation. Technol Forecast Soc Chang 112:262–274
68. Zhao R, Min N, Geng Y, He Y (2017) Allocation of carbon emissions among industries/sectors: an emissions intensity reduction constrained approach. J Clean Prod 142:3083–3094
69. Zhao R, Han JJ, Zhong SZ, Huang Y (2018) Interaction between enterprises and consumers in a market of carbon-labeled products: a game theoretical analysis. Environ Sci Pollut Res 25:1394–1404
70. Zhao R, Geng Y, Liu Y, Tao X, Xue B (2018) Consumers' perception, purchase intention, and willingness to pay for carbon-labeled products: a case study of Chengdu in China. J Clean Prod 171:1664–1671
71. Zheng X, Lin H, Liu Z, Li D, Llopis-Albert C, Zeng S (2018) Manufacturing decisions and government subsidies for electric vehicles in China: a maximal social welfare perspective. Sustainability 10(3):672
72. Zhou KZ, Brown JR, Dev CS (2009) Market orientation, competitive advantage, and performance: a demand-based perspective. J Bus Res 62(11):1063–1070
73. Zhu QH, Sarkis J, Lai KH (2008) Confirmation of a measurement model for green supply chain management practices implementation. Int J Prod Econ 111:261–273
74. Wills R, Finney BA (2004) Environmental systems engineering and economics. Dordrecht: Kluwer Academic Publishers.
75. Zhao R, Neighbour G, McGuire M, Deutz P (2013) A software based simulation for cleaner production: a game between manufacturers and government. J Loss Prevent Proc Indus 26(1): 59-67.
76. Zhao R, Neighbour G, Han J, McGuire M, Deutz P (2012) Using game theory to describe strategy selection for environmental risk and carbon emissions reduction in the green supply chain. J Loss Prevent Proc Indus 25(6): 927-936

Chapter 4
Application of MCDM to Sustainable Materials Selection

4.1 Introduction

Material selection plays a crucial role in product design [9, 44]. A number of studies have emphasized material selection from a "design-oriented" perspective, i.e. from the functional requirements of a product. For example, Chauhan and Vaish [25] applied the VlseKriterijumska Optimizacija I Kompromisno Resenje (VIKOR) and the technique for order preference by similarity to ideal solution (TOPSIS) methods to the classification and selection of soft and hard magnetic materials [25]. Khare et al. [59] proposed a multi-objective optimization model to identify the most suitable phase change materials (PCMs) for latent heat storage design. Similarly, Mansor et al. [75] used the multi-objective optimization methodology for material selection in the design of an automotive brake lever. O'Connor et al. [81] proposed material selection criteria for the design of a new skin-wearable organic solar cell based on its mechanical compliance. Some studies have focused on materials selection based on market demand. For instance, Prasad and Chakraborty [85] used a quality function deployment (QFD)-based approach to select the suitable materials by integrating the voice of the customer with product functionality. Tamani et al. [100] designed a multi-criteria decision support system for fresh food packaging, taking consumer preferences into account. Kwong et al. [61] adopted an artificial intelligence (AI) based methodology for materials selection in design of electric irons, in which the factors of affective design, engineering, and marketing issues were considered. The materials selection is also guided by cost considerations. Rahman et al. [86] developed a decision support system for roof material selection, in which the conceptual cost of the roof element was evaluated at the early stage of building design. Kumar et al. [60] used the entropy with TOPSIS method to evaluate the optimal material for exhaust manifold design, where cost was considered as an important criterion. Ilg et al. [50] calculated the life cycle cost of several high-performance materials (HPM) in infrastructure. Discount rate, expected service life, and external costs were identified as the main factors affecting the performance of the materials.

R. Zhao, *Multi-criteria Decision-Making Approaches to Sustainable Consumption and Production*, Urban Sustainability, https://doi.org/10.1007/978-981-96-2775-2_4

In recent years, low-carbon development has gradually entered product life-cycle management to combat climate change under the umbrella terms such as "green design", "eco-design", "environmentally conscious design" or "design for environment" [15, 21, 40]. Green design, also entitled sustainable design and environmental design is a philosophy that seeks to improve design performance by incorporating environmental, health, and safety attributes into product design decisions, thus reducing the entire life-cycle impacts [5, 26, 46, 77, 89]. The mitigation of the adverse impact of products on the environment and human health can be achieved in part by the careful selection of raw materials, to improve the product lifecycle (including reusability and recyclability), reform the quality of products and minimizing the amount and toxicity of waste generated during manufacturing processes etc. [25, 59]. In this context, the selection of materials should follow the principles of green design to mitigate the negative impact on the environment and human health, enhance product quality and minimize toxicity of materials [14, 123].

This Chapter applies multi-criteria decision making to facilitate material selection, adapt product design to promote sustainable production and consumption, and ultimately achieve long-term commercial success. The Sect. 4.2 introduces a convenient approach to material selection for application to cleaner production, by engineers to aid decision-making taking environmental evaluation into account. The Sect. 4.3 contributes to the literature by developing a computational method to be used by engineers or designers to aid in the selection of commercially available materials by taking typical indicators into account, e.g. economic, environmental, and social performance, thus promoting green design and improving product quality. The Sect. 4.4 introduces a linguistic information based computational approach to aid engineers in the selection of pipe materials. On this basis, the Sect. 4.5 focuses on performing a precise selection of common commercial pipe materials by using a multi-attribute decision making.

4.2 Material Selection for Cleaner Production

In this section, a number of typical indicators by taking environmental criteria into account have been selected for the evaluation. A 'binary-dominance matrix' integrated with 'grey relational analysis' is used to develop an evaluation methodology for application by engineers to aid decision-making in material selection. A spreadsheet based computational tool has been designed to help engineers and designers better understand and apply this approach. Moreover, this work mainly focuses on the selection among similar materials (including type and composition) to further improve the environmental performance of the product. A study to select a range of poly (vinyl chloride) (PVC) materials for the manufacture of handbags is used as a case example.

4.2.1 The Grey Decision Approach

The evaluation approach starts with the selection of the necessary indicators for cleaner production. The use of environmental and economic evaluation indicators is an efficient way to help improve and monitor the processes of product design. According to the basic principles and guidelines of 'Design for Environment' [40, 70, 104], this study has identified a number of typical indicators to represent environmental sustainability as a basis for material selection, e.g. human health risk/toxicity, economic and resource efficiency, in Table 4.1.

As the selected evaluation criteria are not equally important relative to each other and are highly dependent on the product, it is necessary to introduce some form of weighting as part of the evaluation process. Generally, there are two common approaches to determining the extent of weighting required: objective and subjective. The former is mainly based upon the attribute data, e.g. the entropy approach, while the latter is based on the preferences of the decision-maker, e.g. the analytic hierarchy process (AHP) [87, 88, 96]. The 'Binary-Dominance Matrix' method is a simplified way of applying the subjective weighting approach, and has been used in this study, in order to reduce the computational time and complexity [49].

In this approach, all the criteria are listed on both vertical and horizontal axes of a matrix, and then the numerical value 1, 0.5 or 0 is input into each box created by the intersection of the axes according to their relative importance. Each individual criterion is evaluated separately against each other criterion. If a pair of indicators is judged to be of equal importance, a value of 0.5 is assigned to each. Once all the indicators have been assigned a value of 0, 1 or 0.5, each row of the matrix should be increased by 1 to ensure the arithmetic validity. Consequently, the weight can be derived as the result of the aggregated score of each indicator divided by the total score of all the indicators, as shown in the Eq. (4.1).

Table 4.1 Selection of evaluation criteria

Indicators
Energy consumption
Human health risk/toxicity
Materials recyclability
Materials reusability
Use of recycled materials
Manufacture defects
Defects recyclability
By-products
By-products recyclability
Waste production
Manufacturing cost
Sales profit

$$W = \frac{x_i + 1}{\sum_{i=1}^{n} x_i + 1} \tag{4.1}$$

where W is the weight, x_i is the score of the individual criterion.

The grey decision approach is derived from the theory of grey systems and is applied to those uncertain systems with poor or partially known information. General decision models with grey elements, or integrated with grey system models, can be solved using grey analysis [67, 68]. Grey relational analysis is a quantitative method used to aid decision making, by examining how close the value of each attribute is to the value of the ideal parameter. The greater the number of attributes that meet this condition, the greater is the grey relational correlation [120]. In this study, the grey relational analysis is a preferred method for selecting 'environmentally friendly' materials for wider industrial application.

The basic model for grey evaluation can be described as the product of the weight and the result of the grey relational analysis [33], as follows:

$$R = W \times E \tag{4.2}$$

where $R = [r_1, r_2, \ldots, r_m]^T$ is the result vector of m evaluation objects, and W is the weighting value for each evaluation indicator. E is the matrix composed of all the evaluation indicators, and $\delta_i(k)$ is the grey relational coefficient between the kth indicator and the optimal indicator:

$$E = \begin{bmatrix} \delta_1(1) & \delta_1(2) & \cdots & \delta_1(n) \\ \delta_2(1) & \delta_2(2) & \cdots & \delta_2(n) \\ \vdots & \vdots & \vdots & \\ \delta_m(1) & \delta_m(2) & \cdots & \delta_m(n) \end{bmatrix} \tag{4.3}$$

In this study, all the evaluation indicators have been normalized in terms of percentages. It is further assumed that the ideal index of evaluation indicators is $\{\alpha^*\} = [\alpha_1^*, \alpha_2^*, \cdots \alpha_n^*]$, where α_k^* is the optimal value for the kth evaluation indicator. The ideal values are set depending on the target requirements and justification of the manufacturer. Some criteria need to be maximized, such as material recyclability, reusability, sales profit, etc., while others need to be minimized, such as energy consumption, health risk, manufacturing cost, etc. However, it is important to recognize that the ideal value for each criterion should be set at an appropriate level, according to the technical and economic feasibility relevant to a specific manufacturer.

Once α^* has been determined, a new grey relational matrix α can be constructed, where α_k^i is the original numerical value of the kth indicator:

$$\alpha = \begin{bmatrix} \alpha_1^* & \alpha_2^* & \cdots & \alpha_n^* \\ \alpha_1^1 & \alpha_2^1 & \cdots & \alpha_n^1 \\ \vdots & \vdots & & \vdots \\ \alpha_1^m & \alpha_2^m & \cdots & \alpha_n^m \end{bmatrix}. \tag{4.4}$$

Thus, the grey relational coefficient can be calculated as follows:

$$\delta_i(k) = \frac{\underset{i}{min}\underset{k}{min}\left|\alpha_k^* - \alpha_k^i\right| + \varepsilon \underset{i}{max}\underset{k}{max}\left|\alpha_k^* - \alpha_k^i\right|}{\left|\alpha_k^* - \alpha_k^i\right| + \varepsilon \underset{i}{max}\underset{k}{max}\left|\alpha_k^* - \alpha_k^i\right|} \tag{4.5}$$

where ε is known as the distinguishing coefficient and $\varepsilon \in (0,1)$. Empirically, ε is set as 0.5 [67, 68], and this value is adopted in this study.

From Eqs. (4.2) and (4.5), the grey decision-making model can be transformed as follows:

$$R_i = \sum_{k=1}^{n} W(k) \times \delta_i(k) \tag{4.6}$$

The evaluation result R_i can be interpreted as the 'degree of grey incidence', which is used to express the rates of change of the mathematical progressions $\{\alpha^i\}$ relative to an ideal progression $\{\alpha^*\}$. In addition, the evaluation object corresponding to the maximum degree of grey correlation is considered to bethe optimal object, and vice versa.

In order to simplify the computational complexity and to make the grey analysis more convenient for engineers and designers, this approach has been embodied in a computational tool based on a spreadsheet calculation. This tool has been divided into two modules, one for the input parameters and another for the output of the analysis results. An example of the application of this tool is presented in the following section.

4.2.2 An Illustrative Case

A leading handbag manufacturer in Guangdong Province, China, has kindly agreed to use their data being used for this case study. The material commonly used for handbags is PVC, and five different grades with slight, but significant, differences in composition are examined. The designations A, B, C, D and E are used to distinguish the five materials, shown in Fig. 4.1.

The objective is to evaluate these five materials by means of the grey evaluation, and ultimately determine which material is the most sustainable. The evaluation

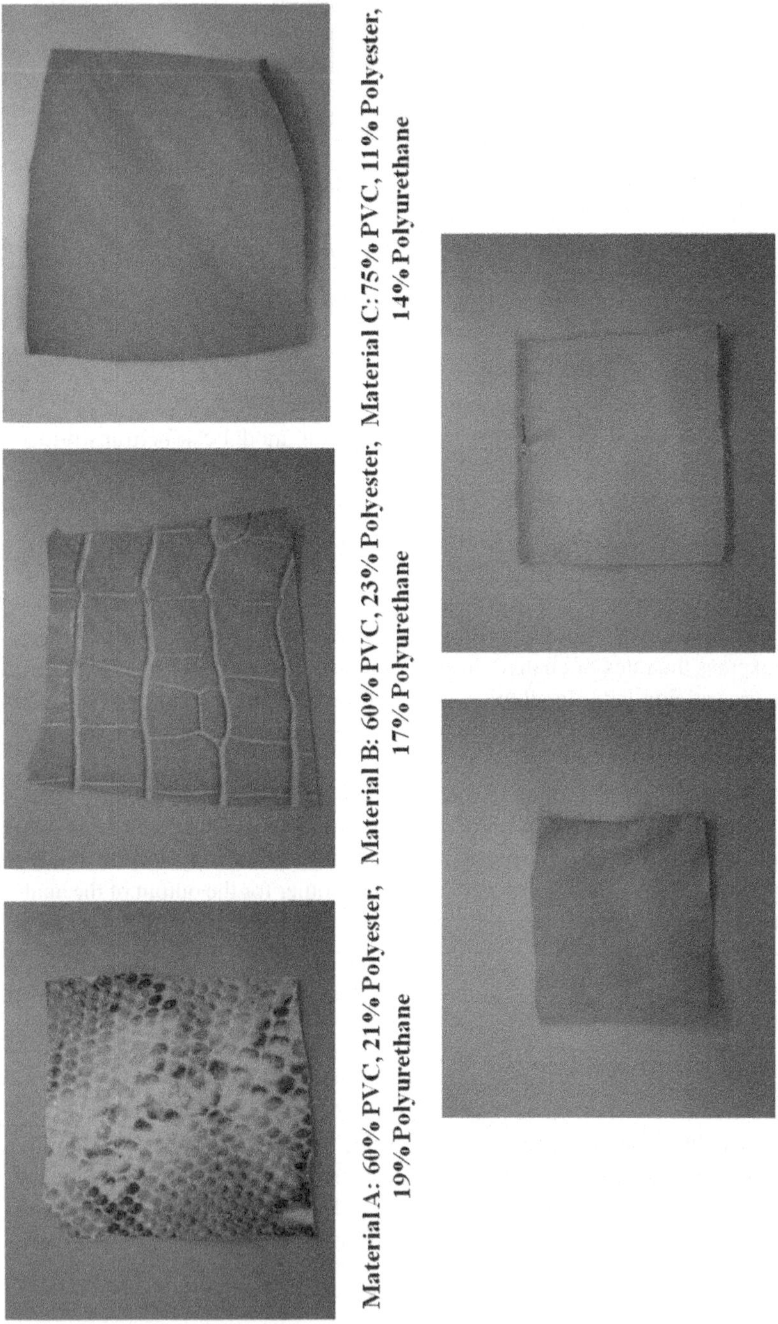

Fig. 4.1 Five PVC materials with different compositions

results are then compared to the current selection of materials used by the cited company. Table 4.2 shows how the selected environmental criteria are calculated.

According to the measurements, the data collected from the handbag manufacturer have been normalized by percentage, as shown in Table 4.3. As for the ideal values, some are set by the production strategies of the cited manufacturer. For example, the manufacturer expects to reduce the power consumption of the PVC material processing to 70% and the manufacturing defects to 2%. Similarly, the manufacturing cost should be minimized to 30%, while the sales profit should be increased to 45%.

In this case example, the product manager of the cited handbag manufacturer was requested to rank the importance of each criterion, which can lead to a more realistic evaluation. The product manager classified the evaluation criteria in seven levels of their relative importance, as shown in Fig. 4.2.

Since some indicators are included in the same level, they are scored 0.5 for each. Based on the importance level, the binary dominance matrix is constructed, as shown in Table 4.4.

Figure 4.3 shows the input module of the developed tool, starting from the weights determination, by marking 0, 0.5 or 1 in the designed binary dominance matrix table (see Fig. 4.3a first step), followed by entering the numerical value for each indicator derived from the case example (see Fig. 4.3b second step).

When all the information has been submitted, the evaluation result can be generated automatically by the spreadsheet programe, and a sequence of the evaluated materials can be ranked from the highest to the lowest grade, as shown in Fig. 4.4. According to the grey evaluation result ®$_E$ > R_C > R_D > R_B > R_A), material E is the optimal material from an 'environmentally friendly' point of view, while material A is the worst choice.

4.2.3 Results

The degree of grey relation 'R_i' reflects how close the data column of each evaluated material is to the selected ideal value (see Table 4.3), and the larger 'R_i' is, the closer the evaluated material is to the ideal value set [67, 68, 120]. According to the result from grey relational analysis ®$_E$ > R_C > R_D > R_B > R_A), material E is the optimal choice in terms of sustainability. This can be deduced from Table 4.3, where the corresponding data show the highest similarity to the ideal values, especially for the criteria of material recyclability, material reusability, by-products and by-products recyclability. On the other hand, material A is shown to be the least sustainable material among them.

The actual consumption of these five PVC materials used by the cited manufacturer has also been examined. Material C has the highest consumption rate for handbag production, followed by materials E, D, B, and A, i.e. $M_C > M_E > M_D > M_B > M_A$, which reflects that the material selection for handbag production is in reality strongly influenced by criteria such as 'market demand', 'materials performance', and 'cost–benefit' [7, 36, 38, 55, 69]. In other words, apart from product

Table 4.2 Definition of measurements for evaluation criteria

Indicators	Measurements
Energy consumption %	$E_c = \frac{E_c(i)}{E_t}$ where $E_c(i)$ is the energy consumed by using the ith material for industrial manufacture, E_t is the total energy consumed by the production, i = A, B, C, D, E. The energy used for PVC materials manufacturing is specified as electricity. For instance, every 100 Kwh electricity, 1 tonne material A will use 97 Kwh for industrial processing, B 93.6 Kwh, C 89.5 Kwh, D 83.0 Kwh and E 80.0 Kwh (Values from Table 4.3)
Human health risk %	$R_h = \frac{R_h(i)}{R_t}$ where $R_h(i)$ is the health risk resulted from using the ith PVC materials for handbags production, i = A, B, C, D, E. $R_h(i) = P_e(i) \times C_e(i)$ where $P_e(i)$ PVC material for industrial manufacture, R_t is the total health risk caused by using all kinds of the PVC materials for handbags production, i = A, B, C, D, E. $R_h(i) = P_e(i) \times C_e(i)$ where $P_e(i)$ is the probability of environmental hazards during selection of the ith material for production, $C_e(i)$ is the consequence of the hazard scenario in terms of the number of fatalities
Materials recyclability %	$R_{ec} = \frac{Q_{rec}(i)}{Q_t(i)}$ where $Q_{rec}(i)$ is the quantity of recycled materials while using the ith PVC material for industrial processing, $Q_t(i)$ is the total quantity of the ith PVC material consuming, i = A, B, C, D, E
Materials Reusability %	$R_{us} = \frac{Q_{us}(i)}{Q_t(i)}$ where $Q_{us}(i)$ is the quantity of reused materials while using the ith PVC material for industrial processing, $Q_t(i)$ is the total quantity of the ith PVC material consuming, i = A, B, C, D, E
Use of recycled materials %	$R_{ur} = \frac{Q_{ur}(i)}{Q_{rec}(i)}$ where $Q_{ur}(i)$ is the usable quantity of the recycled materials, $Q_{rec}(i)$ is the quantity of recycled material while using the ith PVC material for industrial processing, i = A, B, C, D, E
Manufacturing defects %	$M_d = \frac{Q_d(i)}{Q_M(i)}$ where $Q_d(i)$ is the quantity of the manufacture defects, $Q_M(i)$ is the quantity of products (handbags) while using the ith material for industrial manufacturing, i = A, B, C, D, E
Defects recyclability %	$D_{rec} = \frac{Q_{rd}(i)}{Q_d(i)}$ where $Q_{rd}(i)$ is the quantity of the recycled manufacture defects, $Q_d(i)$ is the quantity of the manufacture defects while using the ith material for industrial manufacturing, i = A, B, C, D, E
By-products %	$B_p = \frac{Q_{bp}(i)}{Q_M(i)}$ where $Q_{bp}(i)$ is the quantity of the manufactured by-products, $Q_M(i)$ is the quantity of products (handbags) while using the ith material for industrial manufacturing, i = A, B, C, D, E
By-products recyclability %	$B_{rec} = \frac{Q_{brec}(i)}{Q_{bp}(i)}$ where $Q_{brec}(i)$ is the quantity of the recycled by-products, $Q_{bp}(i)$ is the quantity of the by-products while using the ith material for industrial manufacturing, i = A, B, C, D, E
Waste production %	$W_p = \frac{Q_{wp}(i)}{Q_t(i)}$ where $Q_{wp}(i)$ is the quantity of waste while using the ith material for industrial processing, $Q_t(i)$ is the total quantity of the ith material consuming, i = A, B, C, D, E

(continued)

Table 4.2 (continued)

Indicators	Measurements
Manufacturing cost %	$C_M = \frac{C_M(i)}{C_t}$ where $C_M(i)$ is the cost by using the ith material for industrial production, C_t is the total cost paid by using all kinds of the PVC materials for handbags production, i = A, B, C, D, E
Sales profit %	$S_p = \frac{S_p(i)}{S_t}$ where $S_p(i)$ is the expected sales profit by using the ith material for industrial production, S_t is the total expected sales profit gained by using all kinds of PVC materials for handbags production, i = A, B, C, D, E

Table 4.3 Data for different indicators provided by a handbag manufacturer

Indicators	Material A	Material B	Material C	Material D	Material E	Ideal value
Energy consumption	97.00	93.60	89.50	83.00	80.00	70.00
Human health risk	0.0015	0.0013	0.0005	0.0007	0.0002	0.0001
Materials recyclability	94.70	95.30	94.00	96.30	98.00	98.00
Materials Reusability	0.13	0.16	0.22	0.11	0.23	0.23
Use of recycled materials	90.80	91.00	91.90	92.70	92.00	93.00
Manufacturing defects	8.10	7.30	5.80	6.00	5.00	2.00
Defects recyclability	90.00	93.00	93.60	89.00	94.00	95.00
By-products	0.020	0.017	0.013	0.013	0.011	0.011
By-products recyclability	0.12	0.14	0.16	0.10	0.20	0.20
Waste production	4.90	4.50	3.60	2.90	2.50	2.00
Manufacturing cost	42.00	38.00	32.00	34.00	28.00	30.00
Sales profit	40.00	39.80	38.60	35.70	31.00	45.00

functionality and company profitability, popular fashion is also a fundamental driver in product design. In terms of 'cost–benefit', Table 4.3 shows that material A has the highest manufacturing cost, while it also contributes the highest proportion of profit, both about 40%. Comparatively, the manufacturing cost of material C (about 32%) is lower than that of material A (about 42%), but it yields a higher profit (38.6%). 'Materials performance' is mainly related to the properties of materials that affects the quality of products [7, 55]. For the PVC materials used in the present case, the laboratory performance tests include a mechanical fatigue test, a chemical corrosion test using salt spray, an oven test for thermal expansion and a colour fastness test

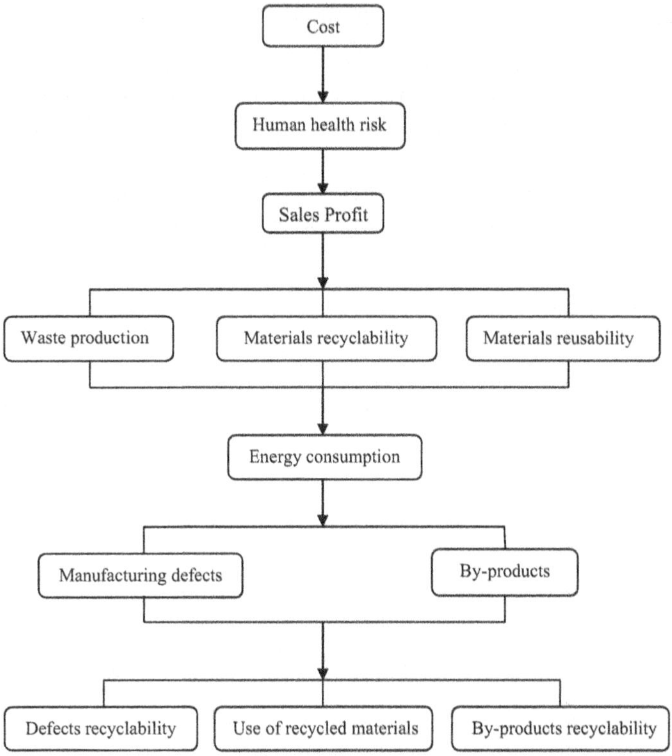

Fig. 4.2 Levels of evaluation criteria ranked by the cited handbag manufacturer

(SATRA Technology 2020, [3, 116]). According to the test results, material C shows the best abrasion resistance, decay resistance, heat resistance, and colour fastness. However, material A shows poor performance, for example, the surface is easily rubbed off during the process of manufacture.

In addition, it is identified that the sequence of actual material consumption ($M_C > M_E > M_D > M_B > M_A$) is quite similar to the result from grey relation analysis ®$_E > R_C > R_D > R_B > R_A$) in the case example. Material C has not only the highest consumption rate for the handbag production, but also the second highest degree of grey incidence. Although material E is superior to C in terms of grey evaluation result, the actual consumption of material E is less than that of C. This is because material C is more favourable as the main material, while material E is more commonly used as an auxiliary material for handbag design. Thus, material C can be determined as the near optimal choice. It is suggested that the actual consumption of material could be integrated into the environmental assessment to help decision making on material selection. In this case study, both materials E and C emerge as 'environmentally friendly' for cleaner production, while the use of material A should be reduced or eliminated all together from the manufacturing process.

Table 4.4 Binary dominance matrix of the case example

Indicators	1	2	3	4	5	6	7	8	9	10	11	12	Score	Weights
1. Energy consumption	×	1	0	1	1	1	1	1	1	0	0	0	7	0.089
2. Human health risk	1	×	1	1	1	1	1	1	1	1	0	1	11	0.139
3. Materials recyclability	1	0	×	0	1	1	1	1	1	0.5	0	0	8	0.101
4. Materials Reusability	1	0	0.5	×	1	1	1	1	1	0.5	0	0	8	0.101
5. Use of recycled materials	0	0	0	1	×	0	0.5	0	0.5	0	0	0	2	0.025
6. Manufacture defects	0	0	0	1	1	×	1	0.5	1	0	0	0	4.5	0.057
7. Defects recyclability	0	0	0	0	0.5	0	×	0	0.5	0	0	0	2	0.025
8. By-products	0	0	0	0	1	0.5	1	×	1	0	0	0	4.5	0.057
9. By-products recyclability	0	0	0	0	0.5	0	0.5	0	×	0	0	0	2	0.025
10. Waste production	1	0	0.5	0.5	1	1	1	1	1	×	0	0	8	0.101
11. Manufacturing cost	1	1	1	1	1	1	1	1	1	1	×	1	12	0.152
12. Sales profit	1	0	1	1	1	1	1	1	1	1	0	×	10	0.127

Previous studies have focused on the selection of materials with different characteristics [13, 20, 37]. The present study confirms that our evaluation methodology is capable of discriminating between materials with similar manufacturing and environmental characteristics. In addition, the compiled calculation tool can be used for a wide range of material selection scenarios.

4.2.4 Summary

This section has introduced a convenient approach for application by designers to aid decision-making in material selection while taking environmental evaluation into account. By integrating the binary dominance matrix with grey relational analysis, a computational tool has been developed using a spreadsheet application. This simplified approach allows for reduced computational cost and ease of use, aimed at encouraging designers to incorporate the concept of sustainability into product design and manufacturing processes. The proposed evaluation approach is flexible, and can provide an environmental index to evaluate a range of materials. This is

	First step: weight determination	Energy consumption %	Human health risk/toxity	Materials Recyclability %	Materials Reusability %	Use of recycled materials %	Manufacture defects %	Defects recyclability %	By-products %	By-products recyclability %	Waste production %	Manufacture cost %	Sales profit %	Total Score	Weight Value
1	Energy consumption %		1	0	0	1	1	1	1	1	0	0	0	7	0.088607595
2	Human health risk/toxity	1		1	1	1	1	1	1	1	1	0	1	11	0.139240506
3	Materials Recyclability %	1	0		0.5	1	1	1	1	1	0.5	0	0	8	0.101265823
4	Materials Reusability %	1	0	0.5		1	1	1	1	1	0.5	0	0	8	0.101265823
5	Use of recycled materials %	0	0	0	0		0	0.5	0	0.5	0	0	0	2	0.025316456
6	Manufacture defects %	0	0	0	0	1		1	0.5	1	0	0	0	4.5	0.056962025
7	Defects recyclability %	0	0	0	0	0.5	0		0	0.5	0	0	0	2	0.025316456
8	By-products %	0	0	0	0	1	0.5	1		1	0	0	0	4.5	0.056962025
9	By-products recyclability %	0	0	0	0	0.5	0	0.5	0		0	0	0	2	0.025316456
10	Waste production %	1	0	0.5	0.5	1	1	1	1	1		0	0	8	0.101265823
11	Manufacture cost %	1	1	1	1	1	1	1	1	1	1		1	12	0.151898734
12	Sales profit %	1	0	1	1	1	1	1	1	1	1	0		10	0.126582278

(a)

	Second step: Parameters Determination	Input name: Material A	Input name: Material B	Input name: Material C	Input name: Material D	Input name: Material E	Ideal Parameters
1	Energy consumption %	97	93.6	89.5	83	80	70
2	Human health risk/toxity	0.0015	0.0013	0.0005	0.0007	0.0002	0.0001
3	Materials Recyclability %	94.7	95.3	94	96.3	98	98
4	Materials Reusability %	0.13	0.16	0.22	0.11	0.23	0.23
5	Use of recycled materials %	90.8	91	91.9	92.7	92	93
6	Manufacture defects %	8.1	7.3	5.8	6	5	2
7	Defects recyclability %	90	93	93.6	89	94	95
8	By-products %	0.02	0.017	0.013	0.013	0.011	0.011
9	By-products recyclability %	0.12	0.14	0.16	0.1	0.2	0.2
10	Waste production %	4.9	4.5	3.6	2.9	2.5	2
11	Manufacture cost %	42	38	32	34	28	30
12	Sales profit %	40	39.8	38.6	35.7	31	45

(b)

Fig. 4.3 Screen shot of the computational tool

Evaluation Result Analysis (Degree of Grey Incidence R_j)		Material A	Material B	Material C	Material D	Material E
1	R_A	0.768372904				
2	R_B		0.795951033			
3	R_C			0.836494293		
4	R_D				0.830072357	
5	R_E					0.860786592

Fig. 4.4 Screen shot of the grey evaluation result

reflected in the case example, where the most sustainable material for manufacturing handbags has been identified from a range of similar PVC materials. It is suggested that the actual consumption of the material could be incorporated into the evaluation process to better justify the material selection in a real situation.

4.3 Materials Selection for Sustainable Design

In order to provide better, safer, and more sustainable products to the public, it is important to predefine the material in the early stages of product design. Materials should be identified for further industrial manufacturing that are more environmentally friendly, e.g. minimize environmental impacts and maximize economic and social benefits. The selection of materials can be influenced by many factors, such as product function, cost, environment, or consumer satisfaction [10, 108]. Because of conflicting or competing objectives, such as economic cost and environmental constraints, the selection of materials for engineering design can be treated as a multi-criteria decision-making problem (MCDM) [12, 48, 88]. MCDM is often divided into two categories, the multiple-objective decision-making (MODM) and multiple-attribute decision-making (MADM) methods [52]. MADM methods such as Analytic Hierarchy Process (AHP), Technique for Order Preference by Similarity to Ideal Solution (TOPSIS), VIekriterijumsko KOmpromisno Rangiranje (VIKOR), and Grey Relational Analysis (GRA) are able to rank material alternatives and help designers determine the best material, defined as the one with the highest degree of satisfaction [22, 23, 43, 51, 53].

The MADM methodology for material selection provides an important decision-making tool when confronted with a number of closely related materials for diverse needs. This study further integrates two common MADM methodologies: AHP, which is used to weight selected indicators, and GRA for further evaluation. GRA is based on the measurement of the degrees of similarity or difference between multiple numerical sequences, which is used to identify the relative influences of the sequences for comparison with a predetermined reference sequence [30, 114]. The GRA indicates the preferred order of various investigated objects in terms of a certain performance index. However, the relative importance of the various indices is difficult to determine, e.g. some of the indicators may be contradictory [112]. In comparison, the AHP is a classic method to fill this gap, as it is used to assess the relative importance among indices. Generally, the purpose of AHP is to decompose a complex decision issue into various hierarchies and factors, which is a typical subjective weighting method based on the judgment and synthesis of decision makers [92]. Thus, an ordered hierarchical decision consists of a dominance relationship in which the weights between different indicators are determined by pairwise comparison [118]. Therefore, the integration of the two methods can lead to a maximization of their advantages, which also facilitates the multi-attribute decision-making of a complex system.

This study gives preferences of selection among similar materials (with similar physical and chemical properties) to investigate their performance. A case example is used to verify the proposed MADM method and demonstrate its practical application. Three alternative polymer materials, i.e. poly (vinyl chloride) (PVC), polypropylene (PP), and polyethylene (PE), are examined to determine their sustainability for plastic pipe design. It is further expected that this integrated approach can help manufacturers to implement sustainable design, and thus improve product sustainability.

4.3.1 Establishment of an Indicator System

As environmental resources are increasingly depleted, products are gradually being developed with less environmental impact, thus moving towards "sustainability." In this context, this study has identified a number of typical indicators to represent product sustainability as a basis for material selection, based on the basic principles and guidelines of sustainable design to construct a decision matrix, as shown in Table 4.5 [32, 40, 109, 123].

Table 4.5 Selected indicators for material selection

Attributes	Indicators
Economic	Direct cost
	Manufacturing cost
	Manufacture defects
	Sales profit
Environmental	Energy consumption
	Health risk
	Waste-gas production
	Wastewater production
	Solid waste production
Social	Product satisfaction
	Material recyclability
	Material reusability
	Use of recycled Materials

For any manufacturer, there is a triple bottom lines (TBL) to run a sustainable business from bottom to top, that is as a basic market demand, alternative materials and energy for production, and the pursuit of societal value for environmental protection [4, 57, 97, 98, 110, 125]. The manufacturer is primarily driven by profit, as economic issues play a dominant role in business operations [4, 34, 45, 78]. Thus, four economic indicators are taken into consideration including direct costs, manufacturing costs, manufacture defects, and sales profit. As the conflicts between economic growth and environmental protection are receiving more and more attention, the most preferred strategy to encompass sustainable development is the implementation of cleaner production [72, 101]. The mitigation of adverse impact of products on the environment and human health can be achieved partly by the careful selection of materials. Thus, five indicators are proposed to represent the product environmental performances: energy consumption, health risk, exhaust gas production, wastewater production, and solid waste production [70, 94].

The social attributes proposed in this study are from the perspective of corporate social responsibility (CSR) to extend the responsibility of manufacturers through the improvement of product sustainability, such as improving product life cycle (including reusability and recyclability), reforming the quality of products, etc., to provide better, safer and sustainable products for public consumption [1, 17, 84]. As proposed by McDonough and Brauggart [115], all materials in a closed-loop social system can be classified into two categories : "technical" or "biological" nutrients. The technical nutrients are those non-toxic and non-harmful materials that can be reused directly in continuous cycles without losing their integrity or quality, and parts of which can be recycled for their quality and functionality, e.g. the cola bottle [18]. In this context, "material reusability", "material recyclability" and "use of recycled materials" are used to reflect the manufacturer's social responsibility for waste prevention.

The use of these evaluation indicators, by considering economic, environmental, and social dimensions, is expected to help improve and monitor the process of product design [39, 91]. The proposed MADM approach combines GRA with AHP to rank alternative materials in terms of their economic, environmental, and social performance. AHP is used to determine the weights of the selected indicators. GRA is then used to develop an evaluation approach to aid material selection. The detailed computational approach is presented below.

4.3.2 Weighting by Analytic Hierarchy Process

Generally, the weighting approach can be divided into two categories: objective weighting and subjective weighting determination [71]. The former obtains the weights derived from the collected data corresponding to each indicator. However, the different indicators may lead to the variation of the weights, and the actual importance of the indicators may be contradictory [2]. The latter mainly depends on the decision maker's preferences regarding the attributes, e.g. AHP is a subjective decision tool that is widely used to determine the weights [117]. Generally speaking, there are three steps in the AHP calculation [105] :

(1) decompose an issue into a hierarchy system: the above attributes are classified into three categories: economic, environmental, and social,
(2) establish a pairwise comparison matrix to compare a set of n attributes in relative importance:

$$A = \begin{bmatrix} a_{11} \cdots a_{1j} \cdots a_{1n} \\ \vdots \quad \vdots \quad \vdots \\ a_{i1} \cdots a_{ij} \cdots a_{in} \\ \vdots \quad \vdots \quad \vdots \\ a_{n1} \cdots a_{nj} \cdots a_{nn} \end{bmatrix} \tag{4.7}$$

where a_1, a_2, ..., a_n, denote different attributes, $a_{ij} = 1/a_{ji}$ and $a_{ij} = a_{ik}/a_{jk}$ indicate how much more important objective i is than objective j to the decision maker. Relative importance is measured on a scale from 1 to 9, where 1 represents that objective i and j are equally important, 3 represents that objective i is only slightly more important than j, 5 represents objective i is somewhat more important than j, 7 represents that objective i is much more important than j, and 9 represents that objective i is far more important than j [92, 117]. We measure the relative importance by using the scale 1–5 in this study.

(3) determine the relative weights; as a_{ij} approximates to w_i/w_j, *i.e.*, $a_{ij} \cong w_i/w_j$, the matrix A can be transformed as follows:

$$W = w_1 \ldots w_j \ldots w_n$$

$$\begin{matrix} w_1 \\ \vdots \\ w_i \\ \vdots \\ w_n \end{matrix} \begin{bmatrix} w_1/w_1 \cdots w_1/w_j \cdots w_1/w_n \\ \vdots \quad \vdots \quad \vdots \\ w_i/w_1 \cdots w_i/w_j \cdots w_i/w_n \\ \vdots \quad \vdots \quad \vdots \\ w_n/w_1 \cdots w_n/w_j \cdots w_n/w_n \end{bmatrix} \tag{4.8}$$

The relative weights are derived by addressing the eigenvector w with the largest eigenvalue $\lambda_{\max}$ of the matrix A, which should satisfy the equation $Aw = \lambda_{\max} w$. In addition, the consistency index $C.I.$ and the consistency ratio $C.R.$, as two inspection indicators, should be calculated to ensure the accuracy of the relative weights and the consistency of the subjective decision. These are:

$$C.I. = (\lambda_{\max} - n)/(n - 1) \tag{4.9}$$

$$C.R. = C.I./R.I. \tag{4.10}$$

where *R.I.* is a random consistency index that varies with the size of the matrices, e.g. 1.11 for a 5 × 5 matrix. *C.I.* should be less than 0.1 to obtain an appropriate result ([92], Tzeng and Huang 2011).

4.3.3 Grey Relational Analysis for Selection

GRA or grey incidence analysis, proposed by Professor Deng Julong in 1984, was developed as a quantitative approach to aid decision-making by examining the degree of similarity or difference between two sequences based on the degree of relationship [29, 30, 68, 106]. The more similar the two sequences are, the higher the grey relational correlation is, and vice versa [30, 67].

To use GRA, it is necessary to normalize the data related to the sequence of selected criteria to a common scale for comparison. Basically, there are three approaches to normalization [30, 114]:

(1) If the ideal value of the sequence of selected criteria is preferred to be larger, such as indicators of sales profit, material recyclability, or reusability, the sequence can be normalized as follows:

$$\beta_i(k) = \frac{\beta_i(k) - min\beta_i(k)}{max\beta_i(k) - min\beta_i(k)} \tag{4.11}$$

where $\beta_i(k)$ is the comparable sequence of selected evaluation criteria.

(2) If the ideal value of the sequence of selected criteria is preferred to be smaller, such as direct cost, manufacturing cost, or energy consumption, the sequence can be normalized as follows:

$$\beta_i(k) = \frac{max\beta_i(k) - \beta_i(k)}{max\beta_i(k) - min\beta_i(k)} \tag{4.12}$$

where $\beta_i(k)$ is the comparable sequence of selected evaluation criteria.

(3) If the ideal value is determined within a numerical value range, the sequence can be normalized as follows:

$$\beta_i(k) = 1 - \frac{|\beta_i(k) - \beta_i|}{max\beta_i(k) - \beta_i} \tag{4.13}$$

where $\beta_i(k)$ is the comparable sequence of selected evaluation criteria, and β_i is the determined ideal value.

Followed by the above normalization, a grey relational coefficient is calculated. It demonstrates the relationship between the dimensionless expected optimal sequence and the original sequence [58, 124]. The grey relational coefficient is:

$$\delta_i(k) = \frac{\underset{i}{min}\underset{j}{min}\|\beta_0(k) - \beta_j(k)\| + \varepsilon \underset{i}{max}\underset{j}{max}\|\beta_0(k) - \beta_j(k)\|}{\|\beta_0(k) - \beta_j(k)\| + \varepsilon \underset{i}{max}\underset{j}{max}\|\beta_0(k) - \beta_j(k)\|} \tag{4.14}$$

where ε is defined as the distinguishing coefficient and $\varepsilon \in (0, 1)$. Generally, ε is set to 0.5 for further calculation [67].

Thus, the degree of grey incidence can be expressed as:

$$\gamma_i = \sum_{k=1}^{n} w_k \cdot \delta_i(k) \tag{4.15}$$

where w_k is the weighting value of the selected indicator k derived from the AHP approach.

The degree of grey incidence indicates the relationship between the comparability sequence and the reference sequence [63, 106]. The reference sequence represents the best performance that can be achieved by any of the comparability sequences. As γ_i becomes much bigger, the comparability sequence approaches the reference sequence. When γ_i reaches the maximum value, a comparability sequence of the corresponding material obtains the highest degree of grey incidence, which means that the alternative material is better than its counterparts. Moreover, the performance of all the alternative materials can be ranked in a sequence, from most optimal to least optimal.

4.3.4 An Illustrative Case

A case example is presented in this section to demonstrate the application of the proposed methodology of material selection for plastic pipe design. The plastic pipes are widely used in the water supply system of Sichuan Province, China, especially for indoor hot and cold water supply. They are made of three alternative materials: poly (vinyl chloride) (PVC), polypropylene (PP), and polyethylene (PE). Here, the PVC is specified as unplasticized PVC (uPVC), with a synthesis of 15% chlorinated polyethylene (CPE) and 75% PVC, and PP as a polypropylene random (PPR), with 5% PE and 95% PP.

In the case example, a set of weight values determined by the chief engineer of the investigated manufacturer and adjusted by the retrieval results of the selected indicators from the online scientific database Scirus is used to avoid subjective judgment in comparing their importance. Scirus.com is a comprehensive academic search engine with more than 440 million scientific items, including published scientific papers,

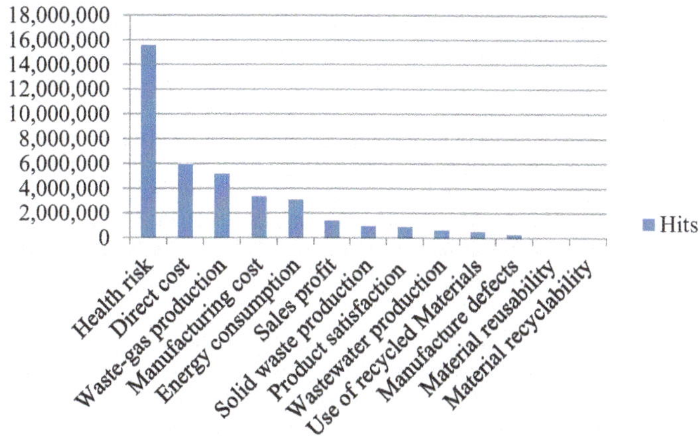

Fig. 4.5 Rank of selected indicators as determined from Scirus.com retrieval

patents, web sources, courseware, and current research progress [122]. Based on the retrieval results of each indicator using "hits" as the measure of importance, the relative importance is ranked from 1 to 13, as shown in Fig. 4.5.

The grey relational result is compared with the actual material selection in the investigated manufacturer of plastic pipe to verify the proposed approach. Table 4.6 shows how the selected indicators are measured for the three alternative materials.

The data for the selected indicators were obtained from a feasibility study report of a plastic pipe manufacturer, located in Leshan City, Sichuan Province, China. According to the measurements, the data have been processed dimensionlessly into percentages for further application of GRA, as shown in Table 4.7.

4.3.5 Results

Based on the relative importance ranking list, the weight value is calculated by the AHP method, as shown in Table 4.8. To simplify the computation, both AHP and GRA have been embodied in a compiled mathematical program implemented by MATLAB. The degree of grey incidence γ_i of the PVC, PP, and PE is calculated as 0.800457709, 0.634945104, and 0.604756842, respectively. Different evaluated alternatives can be ranked in a sequence from the highest to the lowest grade. For instance, the grey incidence ranking is considered the order of superiority for each material in this study. The evaluated alternative material corresponding to the maximum grey correlative degree is considered to be the optimal material, and vice versa. Thus, $\gamma_{PVC} > \gamma_{PP} > \gamma_{PE}$ is obtained, indicating that the PVC is the optimal alternative material and PE is the least optimal choice.

Table 4.8 shows that the indicator "health risk" accounts for about 40% of the weight. This is because the manufacturer's employees are likely to be exposed to toxic

Table 4.6 Measurements of the selected indicators

Indicators	Measurements
Direct cost	$p_i = \frac{\sum_{i=1}^{n} c_i}{C}$ where p_i is the ratio of direct cost of the investigated pipe manufacturer, c_i is the material cost corresponding to the *i*th activity, e.g. $i = 1$ denotes the purchasing cost of a specific material (PVC, PP, PE), $i = 2$ denotes transportation cost of a specific material (PVC, PP, PE), $i = 3$ denotes labor cost of a specific material (PVC, PP, PE), and C is the total investment the investigated pipe manufacturer
Manufacturing cost	$p_m = \frac{c_m}{C}$ where p_m is the ratio of the investigated plastic pipe manufacturing cost, c_m is the plastic pipe manufacturing cost of a specific material (PVC, PP, or PE), and C is the total investment of the pipe manufacturer
Manufacture defects	$p_d = \frac{M_d}{Q_0}$ where p_d is the ratio of the investigated plastic pipe manufacture defects, M_d is the annual quantity of the manufacture defects of the investigated plastic pipe manufacturer, and Q_0 is the annual production of plastic pipes
Sales profit	$p_p = \frac{c_p}{c_s}$ where p_p is the ratio of sales profit of the investigated plastic pipe manufacturer, c_p is the expected sales profit from using a specific material for plastic pipe production (PVC, PP, or PE), and c_s is the sales revenue of the investigated plastic pipe manufacturer
Energy consumption	$p_e = \frac{e_i}{\sum_{i=1}^{n} e_i}$ where p_e is the ratio of energy consumption of the investigated plastic pipe manufacturer, e_i is the energy consumption from using the *i*th fuel, e.g. $i = 1$ denotes the natural gas consumption, $i = 2$ denotes electricity consumption, and $i = 3$ denotes coal consumption
Health risk	Health risk is defined as the potential harm to human health in this study, especially the toxic and poisonous components posed to the workers in the material processing and manufacturing [76, 77] $p_h = \frac{r_i}{\sum_{i=1}^{n} r_i}$ where p_h is the ratio of health risk, r_i is the toxic risk from using the *i*th material for plastic pipe production, i.e. $i = 1$ denotes PVC, $i = 2$ denotes PP, and $i = 3$ PE The toxic risk r_i resulted from using the *i*th material (PP, PE, PVC) is given as follows [78]: $ri = Pi \cdot Ci$ where P_i is the probability of personal injury resulting from the *i*th material process or manufacture, C_i the consequence of the personal injury, such as the number of workers suffered by allergic rhinitis, dermatosis etc.
Waste-gas production	$p_g = \frac{Q_g}{Q_o}$ where p_g is the ratio of waste-gas production of the investigated plastic pipe manufacturer, Q_g is the annual emission of waste gas from using a specific material for production (PVC, PP, or PE), and Q_0 is the annual production of plastic pipes

(continued)

Table 4.6 (continued)

Indicators	Measurements
Wastewater production	$p_W = \frac{Q_W}{Q_o}$ where p_w is the ratio of waste water production of the investigated plastic pipe manufacturer, Q_w is the annual emission of waste water from using a specific material for production (PVC, PP, or PE), and Q_0 is the annual production of plastic pipes
Solid waste production	$P_s = \frac{Q_s}{Q_o}$ where p_s is the ratio of solid waste production of the investigated plastic pipe manufacturer, Q_w is the annual emission of solid waste from using a specific material for production (PVC, PP, or PE), and Q_0 is the annual production of plastic pipes
Product satisfaction	p_z defined as the market share of plastic pipes while using a specific material (PVC, PP, or PE), which is based upon the market of the downstream distribution centers
Material recyclability	Material recyclability is the extent to which waste materials can be converted into reusable material, which are used to make new items, given as follows [79]: $p_r = \frac{Q_r}{Q}$ where Q_r is the theoretical quantity of recycled materials (PVC, PP, or PE), and Q is the total quantity of waste materials produced of the investigated plastic pipe manufacturer
Material reusability	Material reusability is defined as the extent to which the material can be reused directly in continuous cycles without losing their integrity or quality, which is given as follows [80]: $p_R = \frac{Q_R}{Q}$ where Q_R is the theoretical quantity of reused materials (PVC, PP, or PE), and Q is the total quantity of waste materials produced of the investigated plastic pipe manufacturer
Use of recycled materials	$p_t = \frac{Q'}{Q_r}$ where Q' is the actual recycled materials (PVC, PP, or PE), and Q_r is the theoretical quantity of recycled materials of the investigated plastic pipe manufacturer

substances when processing the PVC materials [41]. However, PVC has the highest sales profit and market share (product satisfaction), and the lowest proportion of direct costs and energy consumption. Thus, PVC exhibits the best overall performance among the three alternative materials, even if its inherent health risk is higher than PP or PE. Similarly, PP's degree of grey incidence is higher than that of PE, although PE material is superior to PP on the basis of most of the data for the selected indicators. This is because PP shows better performance in both direct cost and manufacturing cost, which accounts for a large proportion of the relative importance. This shows that the weight value can play a critical role in the sensitivity of the grey relational model in this case study.

Table 4.7 Processed data of the selected indicators

Indicators	Alternative materials		
	PE	PP	PVC
Direct cost	95.5	90.3	90.0
Manufacturing cost	1.43	1.40	10.00
Manufacture defects	1.5	2.0	2.0
Sales profit	29.4	20.0	38.2
Energy consumption	92.5	99.6	39.7
Health risk	0	0	0.00084
Waste-gas production	0.006	1.130	0.230
Wastewater production	11.0	150.0	45.5
Solid waste production	0.09	50.00	0.54
Product satisfaction	25	10	55
Material recyclability	94.50	0.28	65.90
Material reusability	30.7	2.5	16.5
Use of recycled Materials	100	100	98

Table 4.8 Weighting value of the selected indicators based on the Scirus.com retrieval results

Indicators	Weighting value
Direct cost	0.155623917
Manufacturing cost	0.088494826
Manufacture defects	0.008907485
Sales profit	0.038003123
Energy consumption	0.082127819
Health risk	0.404081921
Waste-gas production	0.135173217
Wastewater production	0.018335576
Solid waste production	0.027011282
Product satisfaction	0.024883922
Material recyclability	0.000545566
Material reusability	0.000975515
Use of recycled Materials	0.015835831

In the real world, a manufacturer may define profit as the main criterion of business success, while placing little emphasis on the lifecycle environmental impact of its products [31, 124]. Accordingly, the selection of materials is strongly influenced by market demand, cost–benefit analysis, material properties, and fashion and popularity, as the fundamental drivers of design [123]. For instance, the investigated manufacturer in this study was originally cost-oriented in material selection, and PVC was used for plastic pipe manufacturing. This confirms the evaluation results of the integrated MADM approach. However, a sustainable business should focus on both economic prosperity, environmental and social performance [82]. In green design, the weighting of material selection should pay more attention to the environmental and social perspectives. This could be expressed by increasing the weight values for materials recyclability and reusability to enhance the sustainability of products.

This integrated MADM approach is flexible, which can provide a comprehensive matrix to evaluate a range of alternative materials, including indicators of environmental impact, economic perspective, etc. For instance, a similar study can be found by Zhao et al. [123], who used "binary-dominance matrix" with GRA to aid decision-making on material selection while taking environmental impacts into account. Five PVC materials with close properties were subjected to environmental evaluation to determine their environmental friendliness for handbag manufacturing [123]. Furthermore, Kumar et al. [60] offered a MCDM approach in which the importance of the criteria was calculated using the entropy method, and the alternative materials were ranked by using the TOPSIS method. Wang et al. [112] employed AHP combined with GRA to select a biomass briquette fuel system based on a variety of evaluation criteria, such as economy, cleanliness and environmental protection, production capacity, product quality, and production stability. Yang et al. [119] focused on the material selection from the perspective of remanufacturing, and developed a MCDM model by using the Fuzzy TOPSIS approach to evaluate the performance of the alternative automotive materials. These similar studies have reflected the extension and validity of the proposed method.

However, the weighting plays a critical role in influencing the sensitivity of the GRA, which is subject to the preference of the indicators. The AHP is a subjective weighting method that depends on the judgment and synthesis of the decision makers [6, 102]. It can be replaced by the objective weighting methods, such as the entropy method, TOPSIS, etc., in which the weights are derived from the collected data corresponding to each indicator. In this way, the influence of the variation of the weights on the material selection can be studied to determine the sensitivity of the integrated approach.

4.3.6 Summary

This section presents an integrated multi-attribute decision-making (MADM) approach to support the selection of commercially available materials in the context of sustainable design. The MADM couples grey relational analysis (GRA) with an analytic hierarchy process (AHP) to rank alternative materials in terms of their economic, environmental, and social performance. AHP is used to examine the relative importance and assign the corresponding weights for each indicator, and GRA is applied to rank the alternative materials based on the multiple attributes of their performance. In addition, a case example is used to verify the proposed MADM method and demonstrate its practical application. Three alternative polymer materials, i.e. poly (vinyl chloride) (PVC), polypropylene (PP), and polyethylene (PE), are investigated to determine their sustainability for plastic pipe design. With the weights for the selected criteria being determined by the chief engineer of the investigated manufacturer and adjusted by keyword retrieval from the Scirus database, PVC is determined to be the optimal material, and PE the least optimal.

4.4 Preference Based Decision Making for Pipe Materials Selection

Engineering designers usually express their preferences or judgments in a linguistic environment, which can lead to ambiguity [73]. This section introduces a multi-criteria decision making (MCDM) approach that couples the Inter-valued Trapezoidal Intuitionistic Fuzzy Number (IVTIFN) with the Technique for Order Preference by Similarity to Ideal Solution (TOPSIS) to facilitate the selection of pipe materials. The TOPSIS is unable to deal with such vagueness, which calls for further improvement [93]. Fuzzy set provides the insight into the development of TOPSISby transforming the linguistic information into fuzzy numbers, to indicate the preference and judgment of the decision maker [65, 113]. However, the linguistic variables in an ordinary fuzzy set are not clear enough to express the decision maker's preference by using an exact numerical value of either 0 or 1 [35]. In such a case, this study applies IVTIFN to handle with ambiguity in linguistic information, by which its predetermined numerical interval can better specify the fuzzification in a fixed bounded limit [111].

Their integration can maximize the advantage in better expressing the decision maker's preference on the proposed evaluation criteria by using a bounded limit instead of an exact value, to rank material alternatives based on their functional, economic and environmental attributes. To reduce the possible information overlap resulting from the criteria, Mahalanobis Distance is incorporated into IVTIFN-TOPSIS to discriminate the similarities of alternative materials by eliminating the highly correlated decision criteria. An illustrative example is provided to verify the proposed approach and demonstrate its practical application, where four common alternative materials, including carbon steel, galvanized steel, polyvinyl chloride (PVC), and high density polyethylenes (HDPE), are subjected to precise selection to determine their adaptability in wastewater piping. The study is expected to provide insight into sustainable design of wastewater piping system, ultimately improving their sustainability.

4.4.1 Formulation of an Indicator System

An indicator system for piping material selection is established, as shown in Table 4.9. These indicators are divided into three categories, indicating the functional, economic and environmental attributes of materials, to reflect the most crucial criteria that should be considered in materials selection for a piping system design.

The functional attribute reflects the specific performance that a particular material must meet [17]. In this study, a number of typical criteria are identified to examine the functional performance of pipe materials, in order to ensure reliability and reduce potential environmental risks during wastewater transport, including tensile strength (C1), elastic modulus (C2), linear expansion (C3), scaling resistance (C4) and corrosion resistance (C5). C1 reflects the ability of materials to withstand the maximum load of tensile stress [62]. C2 reflects the ability of the materials to resist elastic deformation, while C3 shows material's resistance to deformation caused by temperature change [90, 103]. Scaling and corrosion resistance play key a role in the functional performance of pipe materials to decrease pipe failure [54].

The cost plays an important role in the material selection, which is considered by three criteria, containing the marketing price of materials (C6), density (C7) and Hazen-Williams roughness (C8). It should be noted that C7 and C8 have an indirect impact on the cost of pipeline construction and operation. The pipe material with higher density may cause heavier cost of pipeline installation [121]. In addition, the degree of roughness of the pipe material is positively correlated with the maintenance cost in swage transportation [27].

With the gradual emergence of environmental considerations in product design, selection of materials also focuses on the environmental impact throughout the life cycle [123]. In this context, three indicators are proposed to indicate the environmental performance of pipe materials: energy consumption (C9), human health risk (C10), and material recyclability (C11). Energy consumption here focuses on the power consumption related to pipeline construction and maintenance, which may contribute to indirect carbon emissions [123]. Health risk indicates the hazard of toxic components released to the workers during the processing and welding of pipe materials [126]. The material recyclability is used to reflect the potential of waste prevention, thus reducing landfill disposal of waste materials [127].

4.4.2 Model Construction

In this study, the data corresponding to the performance of alternatives and preferences on the criteria (shown in Table 4.9) are mainly determined by engineering designers, which are generally presented in a linguistic way. Thus, IVTIFN is employed to convert such raw data into fuzzy numbers.

Table 4.9 An indicator system of pipe materials selection

Indicator	Criteria		Implication
Functional property (P1)	Tensile strength	C1	Tensile strength is the maximum load of tensile stress that material could endure. An excellent pipe material should withstand heavy force
	Elastic modulus	C2	Elastic modulus reflects the ability that material resists elastic deformation
	Linear expansion	C3	Linear expansion represents the relationship between line size and temperature, reflecting material's resistibility to deformation caused by temperature changes
	Corrosion resistance	C4	Corrosion resistance reflects the ability that pipe material bears the chemical corrosion caused by waste water
	Scaling resistance	C5	Scaling resistance indicates the pipe material's ability of anti-clogging during waste water transport
Economic property (P2)	Marketing price	C6	In this study, the retailing price is to reflect materials marketing price
	Density	C7	The pipe material with a larger density may bring a heavier cost on pipeline installation
	Hazen-Williams roughness	C8	Hazen-Williams roughness is a common used parameter to measure head loss in water distribution
Environmental property (P3)	Energy consumption	C9	Energy consumption specifies electricity consumption during the materials production
	Health risk	C10	The hazard of toxic components posed to the employees in processing and welding of pipe materials
	Recyclability	C11	The potential of waste materials can be recycled

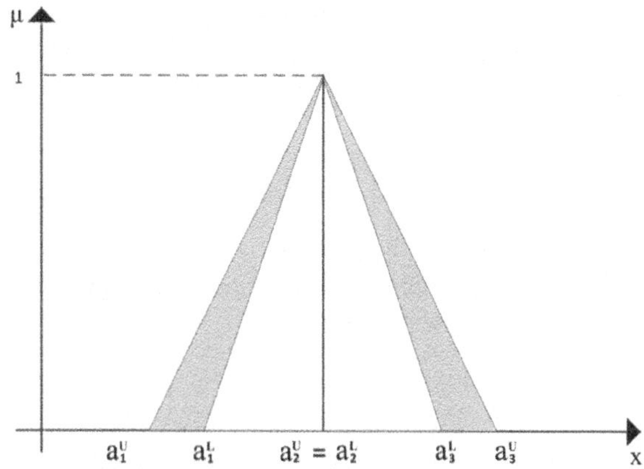

Fig. 4.6 Membership function of IVTIFN

A set of IVTFN (shown in Fig. 4.6) is defined as $\tilde{A} = \left(\tilde{A}^{L}, \tilde{A}^{U}\right) = \left[\left(a_1^L, a_2^L, a_3^L\right); \left(a_1^U, a_2^U, a_3^U\right)\right]$, ($a_2^L = a_2^U = a_2$). The membership function ($\mu_{\tilde{A}^L}(x)$, $\mu_{\tilde{A}^U}(x)$), indicating the degree to which the element (x) is subordinate to the set, is expressed as follows [16]:

$$\mu_{\tilde{A}^L}(x) = \begin{cases} \frac{x-a_1^L}{a_1^L-a_2^L}, x \in \left[a_1^L, a_2^L\right] \\ \frac{a_3^L-x}{a_3^L-a_2^L}, x \in \left[a_2^L, a_3^L\right] \\ 0, \textit{Otherwise} \end{cases}, \mu_{\tilde{A}^U}(x) = \begin{cases} \frac{x-a_1^U}{a_1^U-a_2^U}, x \in \left[a_1^L, a_2^L\right] \\ \frac{a_3^U-x}{a_3^U-a_2^U}, x \in \left[a_2^L, a_3^L\right] \\ 0, \textit{Otherwise} \end{cases} \tag{4.16}$$

The arithmetic operation for the two fuzzy numbers is given as follows:

$$\begin{aligned} \tilde{\alpha} + \tilde{\beta} &= \left[\left(\alpha_1^L, \alpha_2^L, \alpha_3^L\right); \left(\alpha_1^U, \alpha_2^U, \alpha_3^U\right)\right] + \left[\left(\beta_1^L, \beta_2^L, \beta_3^L\right); \left(\beta_1^U, \beta_2^U, \beta_3^U\right)\right] \\ &= \left[\left(\alpha_1^L + \beta_1^L, \alpha_2^L + \beta_2^L, \alpha_3^L + \beta_3^L\right); \left(\alpha_1^U + \beta_1^U, \alpha_2^U + \beta_2^U, \alpha_3^U + \beta_3^U\right)\right] \end{aligned} \tag{4.17}$$

$$\begin{aligned} \tilde{\alpha} - \tilde{\beta} &= \left[\left(\alpha_1^L, \alpha_2^L, \alpha_3^L\right); \left(\alpha_1^U, \alpha_2^U, \alpha_3^U\right)\right] - \left[\left(\beta_1^L, \beta_2^L, \beta_3^L\right); \left(\beta_1^U, \beta_2^U, \beta_3^U\right)\right] \\ &= \left[\left(\alpha_1^L - \beta_1^L, \alpha_2^L - \beta_2^L, \alpha_3^L - \beta_3^L\right); \left(\alpha_1^U - \beta_1^U, \alpha_2^U - \beta_2^U, \alpha_3^U - \beta_3^U\right)\right] \end{aligned} \tag{4.18}$$

$$\begin{aligned} \tilde{\alpha} \times \tilde{\beta} &= \left[\left(\alpha_1^L, \alpha_2^L, \alpha_3^L\right); \left(\alpha_1^U, \alpha_2^U, \alpha_3^U\right)\right] \times \left[\left(\beta_1^L, \beta_2^L, \beta_3^L\right); \left(\beta_1^U, \beta_2^U, \beta_3^U\right)\right] \\ &= \left[\left(\alpha_1^L\beta_1^L, \alpha_2^L\beta_2^L, \alpha_3^L\beta_3^L\right); \left(\alpha_1^U\beta_1^U, \alpha_2^U\beta_2^U, \alpha_3^U\beta_3^U\right)\right] \end{aligned} \tag{4.19}$$

$$\begin{aligned} \tilde{\alpha} \div \tilde{\beta} &= \left[\left(\alpha_1^L, \alpha_2^L, \alpha_3^L\right); \left(\alpha_1^U, \alpha_2^U, \alpha_3^U\right)\right] \div \left[\left(\beta_1^L, \beta_2^L, \beta_3^L\right); \left(\beta_1^U, \beta_2^U, \beta_3^U\right)\right] \\ &= \left[\left(\alpha_1^L/\beta_1^L, \alpha_2^L/\beta_2^L, \alpha_3^L/\beta_3^L\right); \left(\alpha_1^U/\beta_1^U, \alpha_2^U/\beta_2^U, \alpha_3^U/\beta_3^U\right)\right] \end{aligned} \tag{4.20}$$

Assume that the number of engineering designers involved in the pipe material selection is e, and the kth designer's fuzzy rating corresponding to the jth alternative material in the ith criterion is $\tilde{r}_{ijk} = \left[\left(r^L_{ijk1}, r^L_{ijk2}, r^L_{ijk3}\right), \left(r^U_{ijk1}, r^U_{ijk2}, r^U_{ijk3}\right)\right]$. Thus, the fuzzy ratings of different designers are aggregated as $\tilde{r}_{ij}$, given as follows:

$$\tilde{r}_{ij} = \left[\left(r^L_{ij1}, r^L_{ij2}, r^L_{ij3}\right), \left(r^U_{ij1}, r^U_{ij2}, r^U_{ij3}\right)\right] = \frac{1}{e}\sum_{k=1}^{e}\tilde{r}_{ijk} \quad (4.21)$$

The obtained aggregated ratings need to be normalized, and the normalization procedures are given as follows:

$$\begin{aligned}\tilde{x}_{ij} &= \left[\left(x^L_{ij1}, r^L_{ij2}, x^L_{ij3}\right), \left(x^U_{ij1}, x^U_{ij2}, x^U_{ij3}\right)\right] \\ &= \left[\left(\frac{r^L_{ij1}}{r^U_{j3}}, \frac{r^L_{ij2}}{r^U_{j3}}, \frac{r^L_{ij3}}{r^U_{j3}}\right), \left(\frac{r^U_{ij1}}{r^U_{j3}}, \frac{r^U_{ij2}}{r^U_{j3}}, \frac{r^U_{ij3}}{r^U_{j3}}\right)\right], j \in B\end{aligned} \quad (4.22)$$

$$\begin{aligned}\tilde{x}_{ij} &= \left[\left(x^L_{ij1}, x^L_{ij2}, x^L_{ij3}\right), \left(x^U_{ij1}, x^U_{ij2}, x^U_{ij3}\right)\right] \\ &= \left[\left(\frac{r^L_{j1}}{r^U_{ij3}}, \frac{r^L_{j1}}{r^U_{ij2}}, \frac{r^L_{j1}}{r^U_{ij1}}\right), \left(\frac{r^L_{j1}}{r^L_{ij3}}, \frac{r^L_{j1}}{r^U_{ij2}}, \frac{r^L_{j1}}{r^L_{ij1}}\right)\right], j \in C\end{aligned} \quad (4.23)$$

where B denotes the beneficial criteria, $r^U_{j3} = \max_i r^U_{ij3}$; whilst C represents cost criteria, $r^L_{j1} = \min_i r^L_{ij1}$.

The best and worst value of the jth criterion is denoted as $\tilde{x}^*_j = \max_i \tilde{x}_{ij}$, $\tilde{x}^-_j = \min_i \tilde{x}_{ij}$, respectively. The normalized rating is:

$$\tilde{a}_{ij} = \frac{\tilde{x}_{ij} - \tilde{x}^-_j}{\tilde{x}^*_j - \tilde{x}^-_j} \quad (4.24)$$

The normalized matrix of ratings is thus constructed:

$$A = \begin{bmatrix} \tilde{a}_{11} & \cdots & \tilde{a}_{1n} \\ \vdots & \ddots & \vdots \\ \tilde{a}_{m1} & \cdots & \tilde{a}_{mn} \end{bmatrix} \quad (4.25)$$

where m indicates the number of material alternatives, n is the number of evaluation criteria.

The engineering designers can use linguistic remarks, including Excellent (E), Good (G), Fair (F), Poor (P), Very Poor (VP), to classify the performance of alternatives with respect to each criterion. This study determines the alternative materials performance by analogy with such remarks, and converts them into corresponding fuzzy numbers, as shown in Table 4.10.

Table 4.10 Transformation of linguistic remarks to fuzzy numbers

Linguistic remark		Fuzzy number
Performances of alternatives	Relative importance of criteria	
Very poor (VP)	Extremely low (EL)	((0.1, 0.1, 0.2), (0.05,0.1, 0.25))
Poor (P)	Low (L)	((0.2, 0.3, 0.4), (0.15, 0.3, 0.45))
Fair (F)	Medium (M)	((0.4, 0.5,0.6), (0.35, 0.5,0.65))
Good (G)	High (H)	((0.6, 0.7,0.8), (0.55, 0.7, 0.85))
Excellent (E)	Extremely high (EH)	((0.8, 0.9, 0.9), (0.75, 0.9, 0.95))

Assume the weightings of the jth criteria assessed by the kth designer is $\tilde{z}_{jk} = \left[\left(z^L_{jk1}, z^L_{jk2}, z^L_{jk3}\right), \left(z^U_{jk1}, z^U_{jk2}, z^U_{jk3}\right)\right]$. The obtained fuzzy weights are aggregated as follows:

$$\tilde{z}_j = \left[\left(z^L_{j1}, z^L_{j2}, z^L_{j3}\right), \left(z^U_{j1}, z^U_{j2}, z^U_{j3}\right)\right] = \frac{1}{e}\sum\nolimits_{k=1}^{e} \tilde{z}_{jk} \tag{4.26}$$

To facilitate TOPSIS application, the obtained weights shown in Eq. (4.25) have to be normalized as follows:

$$\tilde{w}_j = \frac{\tilde{z}_j}{\sum_j \tilde{z}_j} \tag{4.27}$$

Thus, the weighting matrix is constructed as follows:

$$w = \mathrm{diag}\left(\tilde{w}_1, \tilde{w}_2, \ldots, \tilde{w}_n\right) \tag{4.28}$$

The above normalized ratings of material alternatives and their associated weights are input into TOPSIS to obtain the distances between alternatives and the best/least alternative. These distances are based on Euclidean Distance (IVTIFN-TOPSISE) and Mahalanobis Distance (IVTFN-TOPSISM) respectively, to calculate their closeness coefficients for ranking material alternatives.

The Euclidean Distances between the material alternatives and the best/least alternative are specified as follows:

$$\tilde{D}^{+}_{Ei} = \sqrt{\sum_j \left[\tilde{w}_j\left(\tilde{a}_{ij} - \tilde{a}^{*}_j\right)\right]^2} \tag{4.29}$$

$$\tilde{D}^{-}_{Ei} = \sqrt{\sum_j \left[\tilde{w}_j\left(\tilde{a}_{ij} - \tilde{a}^{-}_j\right)\right]^2} \tag{4.30}$$

The closeness coefficient is:

$$\tilde{C}_{Ei} = \frac{\tilde{D}^-_{Ei}}{\tilde{D}^-_{Ei} + \tilde{D}^+_{Ei}} \tag{4.31}$$

where $\tilde{C}_{Ei} = \left[\left(c^L_{Ei1}, c^L_{Ei2}, c^L_{Ei3}\right), \left(c^U_{Ei1}, c^U_{Ei2}, c^U_{Ei3}\right)\right]$

A defuzzification method, Centre of Gravity (COG), is applied to convert fuzzy numbers $\tilde{C}_{Ei}$ into crisp scores C_{Ei} to rank the alternatives, given as follows [19]:

$$C_{Ei} = \sqrt{C^L_{Ei} \times C^U_{Ei}} = \sqrt{\frac{\int \mu^L_{\tilde{C}_{Ei}}(x)xdx}{\int \mu^L_{\tilde{C}_{Ei}}(x)dx} \times \frac{\int \mu^U_{\tilde{C}_{Ei}}(x)xdx}{\int \mu^U_{\tilde{C}_{Ei}}(x)dx}} \tag{4.32}$$

The greater C_{Ei} is, the better the corresponding material is.

The Mahalanobis Distances between alternatives and the best/least alternative are:

$$\tilde{D}^+_{Mi} = \sqrt{(A_i - A^*)^T w^T \Sigma^{-1} w (A_i - A^*)} \tag{4.33}$$

$$\tilde{D}^-_{Mi} = \sqrt{\left(A_i - A^-\right)^T w^T \Sigma^{-1} w \left(A_i - A^-\right)} \tag{4.34}$$

where A_i A^*, and A^- indicates the ith material alternative, the best alternative and the least alternative, respectively, $A_i = \left(\tilde{a}_{i1}, \tilde{a}_{i2}, \ldots, \tilde{a}_{in}\right)^T$, $A^* = \left(\tilde{a}^*_1, \tilde{a}^*_2, \ldots, \tilde{a}^*_n\right)^T$, $A^- = \left(\tilde{a}^-_1, \tilde{a}^-_2, \ldots, \tilde{a}^-_n\right)^T$ ($\tilde{a}^*_j = \max_i \tilde{a}_{ij}$, $\tilde{a}^-_j = \min_i \tilde{a}_{ij}$). Σ is the co-variance matrix of A. If Σ is indicated as a singular matrix, the Moore–Penrose generalized inverse matrix Σ^{-1} is used to replace Σ [24].

The closeness coefficient is:

$$\tilde{C}_{Mi} = \frac{\tilde{D}^-_{Mi}}{\tilde{D}^-_{Mi} + \tilde{D}^+_{Mi}} \tag{4.35}$$

where $\tilde{C}_{Mi} = \left[\left(c^L_{Mi1}, c^L_{Mi2}, c^L_{Mi3}\right), \left(c^U_{Mi1}, c^U_{Mi2}, c^U_{Mi3}\right)\right]$

Similarly, the fuzzy numbers $\tilde{C}_{Mi}$ is converted into crisp scores C_{Mi} to rank the alternatives. The greater C_{Mi} is, the better the corresponding material is.

4.4.3 An Illustrative Case

The case example is to conduct material selection for waste water piping transport in a newly constructed municipal sewage treatment plant, Chengdu City, China. According to the design requirements, the nominal diameter of the pipeline is 200 mm, which must withstand the pressure of 1.6 MPa. Currently, there are a wide variety of commercial pipe materials in the market, including carbon steel, copper, ductile iron, Polyethylene (PE), Polyvinyl chloride (PVC), Pentatricopeptide repeats (PPR), etc. Given that their applications in sewage treatment, this study chooses four common pipe materials, i.e. carbon steel (M1), galvanized steel (M2), PVC (M3) and HDPE (M4) for further precise selection.

A group comprised of 5 experienced engineers were involved in the selection of pipe materials, including 2 project managers from the landfill site and 3 engineers involved in the design of the leachate collection and treatment system. Their subjective judgments on the performance of these four alternative materials for each criterion are given in Table 4.11. Table 4.12 shows the subjective judgments of the five engineers involved on the importance of the proposed criteria.

Table 4.11 Subjective judgments on performances of the alternative materials

		C1	C2	C3	C4	C5	C6	C7	C8	C9	C10	C11
Engineer 1	M1	E	E	VP	VP	VP	G	G	F	G	G	G
	M2	E	E	VP	P	P	E	G	F	E	E	E
	M3	F	F	F	G	G	P	P	E	P	P	P
	M4	P	P	E	E	E	F	VP	G	P	VP	VP
Engineer 2	M1	E	E	F	P	P	E	E	F	E	E	G
	M2	E	E	F	P	P	E	E	F	E	E	G
	M3	F	F	G	G	G	P	P	E	P	P	F
	M4	VP	VP	E	E	E	F	VP	G	P	P	P
Engineer 3	M1	E	E	P	P	VP	F	G	F	G	G	G
	M2	P	VP	G	G	G	P	P	G	P	F	P
	M3	F	F	E	E	E	F	P	G	VP	P	VP
	M4	E	E	VP	VP	VP	G	G	F	G	G	G
Engineer 4	M1	E	E	VP	P	P	E	G	F	E	E	E
	M2	F	F	F	G	G	P	P	E	P	P	P
	M3	P	P	E	E	E	F	VP	G	P	VP	VP
	M4	E	E	F	P	P	E	E	F	E	E	G
Engineer 5	M1	E	E	F	P	P	E	E	F	E	E	G
	M2	F	F	G	G	G	P	P	E	P	P	F
	M3	VP	VP	E	E	E	F	VP	G	P	P	P
	M4	E	E	P	P	VP	F	G	F	G	G	G

Table 4.12 Importance of criteria

Criteria		Engineer 1	Engineer 2	Engineer 3	Engineer 4	Engineer 5
Functional attribute	C1	EL	L	L	L	EL
	C2	EL	L	EL	L	EL
	C3	EL	EL	EL	EL	EL
	C4	EH	H	H	H	EH
	C5	EH	EH	H	H	H
Economic attribute	C6	EH	EH	EH	EH	EH
	C7	L	M	L	L	L
	C8	L	M	M	L	L
Environmental attribute	C9	EL	EL	EL	EL	EL
	C10	EL	EL	L	L	L
	C11	EL	EL	L	L	L

4.4.4 Results

Table 4.13 shows the ranking result of the IVTIFN-TOPSIS(E) model, which is M3 > M4 > M1 > M2. The ranking results reflect that two optional plastic materials, i.e. PVC (M3) and HDPE (M4), are the optimal and near-optimal materials. Figure 4.7 shows that they have excellent performance in most of the criteria. For instance, PVC and HDPE show better capacities in corrosion resistance (C4) and scaling resistance (C5), which are the main functional requirements to ensure the liability of the piping system. In contrast to M3 and M4, they have their own advantages and disadvantages. PVC shows superiority in mechanical properties (C1 and C2), marketing price (C6) and roughness (C8). Nevertheless, HDPE is more environmentally friendly and capable of anti-corrosion. Taking the relative importance of the criteria into account, the most suitable material for leachate pipelines is PVC (M3). M1 performs better than M2 in most of the evaluation criteria, except for C4 and C5.

A correlation test is performed to examine the correlation between criteria using Pearson's analysis, through which Table 4.14 shows that a number of criteria are significantly correlated with each other, highlighted by the functional attributes, i.e. between C1–C5.

The major purpose of the material selection is to ensure reliable piping of wastewater, where anti-corrosion and anti-clogging are the prerequisites. The criteria C4 and C5 have directly reflected such functional performances. Although C1, C2 and C3

Table 4.13 Results of TOPSIS(E) model

Model	M1	M2		M3	M4
IVTFN-TOPSIS(E)	C_{Ei}	0.3272	0.2185	0.7170	0.6426
	Ranking order	3	4	1	2

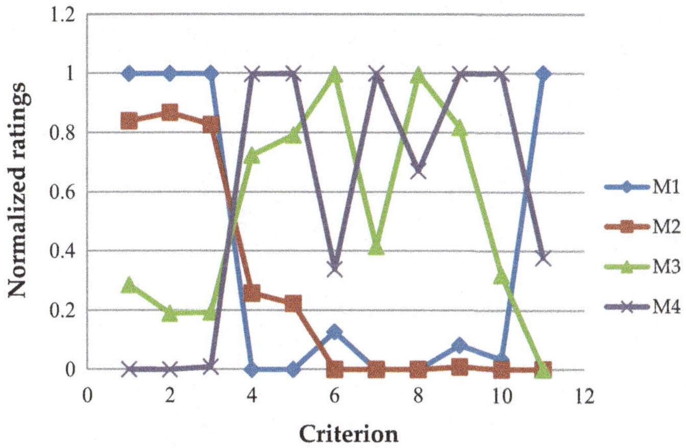

Fig. 4.7 Performance comparisons among the four alternative materials

have certain relationship linked with pipe reliability, their performance in wastewater piping is of lower priority in contrast to C4 and C5. In this context, they are eliminated from the 11 criteria to reduce information overlap. The Mahalanobis distance is further employed to check if the ranking results are different.

Table 4.15 shows that the materials alternative ranking by using IVTIFN-TOPSIS(M) is M4 > M3 > M2 > M1. Compared with the result by using the IVTIFN-TOPSIS(E), it is clear that HDPE is regarded as the optimal material, while carbon steel is the least option.

Both the results of the IVTIFN-TOPSIS(E) and the IVTIFN-TOPSIS(M) model indicate that polymers are superior to the metallic materials, which is consistent with that of Petit-Boix et al. [83] indicating plastic materials are suitable for sewage transport due to their excellent performance on anti-corrosion. Zhao et al. [126] combined AHP with gray relational analysis to select similar plastic pipes for heating system, through which PVC showed great advantage in cost saving and PE was better in its performance on anti-clogging. Li [64] further identified that HDPE was widely used in landfill leachate transport due to its receptivity to highly concentrated organic matter. Such a result may validate the rationality of the ranking results. Anojkumar et al. [129] further identified that there was no significant difference in the application of TOPSIS and VIKOR for pipe material selection, but the former is much simpler and time-saving in computation. This may reflect the feasibility of our proposed approach.

Based on the subjective judgments of the engineers interviewed, they attached great importance to the price, corrosion resistance and scaling resistance when selecting thepipe materials. This focus indicates that cost and functional properties are still the most important factors in material selection [11]. However, the engineers pay little attention to the environmental impacts of materials, indicating that environmental management is still not taken as a significant criterion in engineering design

Table 4.14 Correlation matrix of the evaluation criteria

Criteria	C1	C2	C3	C4	C5	C6	C7	C8	C9	C10	C11
C1	1	0.9949	0.9951	− 0.996	− 0.996	− 0.576	− 0.952	− 0.854	− 0.974	− 0.907	0.4032
C2	0.9949	1	0.9996	− 0.987	− 0.996	− 0.655	− 0.924	− 0.902	− 0.985	− 0.87	0.4158
C3	0.9951	0.9996	1	− 0.99	− 0.998	− 0.652	− 0.92	− 0.9	− 0.98	− 0.864	0.4396
C4	− 0.996	− 0.987	− 0.99	1	0.9966	0.5508	0.9364	0.8357	0.9499	0.8906	− 0.476
C5	− 0.996	− 0.996	− 0.998	0.9966	1	0.6165	0.9224	0.8779	0.9665	0.869	− 0.473
C6	− 0.576	− 0.655	− 0.652	0.5508	0.6165	1	0.3526	0.9172	0.6853	0.2399	− 0.407
C7	− 0.952	− 0.924	− 0.92	0.9364	0.9224	0.3526	1	0.6882	0.9967	0.9922	− 0.16
C8	− 0.854	− 0.902	− 0.9	0.8357	0.8779	0.9172	0.6882	1	0.9108	0.5947	− 0.455
C9	− 0.974	− 0.985	− 0.98	0.9499	0.9665	0.6853	0.9967	0.9108	1	0.8713	− 0.274
C10	− 0.907	− 0.87	− 0.864	0.8906	0.869	0.2399	0.9922	0.5947	0.8713	1	− 0.077
C11	0.4032	0.4158	0.4396	− 0.476	− 0.473	− 0.407	− 0.16	− 0.455	− 0.274	− 0.077	1

Table 4.15 Results of TOPSIS(M) model

Model	M1	M2		M3	M4
IVTFN-TOPSIS(M)	C_{Mi}	0.1461	0.3269	0.6430	0.7094
	Ranking order	4	3	2	1

[123]. With the increasing extraction of raw materials , engineering design needs to consider the transition to sustainability, thus not only to follow the economic bottom line, but also to improve the environmental performance [42].

4.4.5 Summary

In this study, IVTIFN-TOPSIS is used to select materials for sewer pipes, where IVTIFN is used to deal with subjective and linguistic information. To overcome information overlap and overestimation of evaluation criteria, Mahalanobis Distance is introduced to improve the computational approach by analyzing the correlations among the criteria. As a case example, the application of the model is verified in the selection of materials for the pipes of a local municipal wastewater treatment plant. In addition to the general attentions given to the materials functionalities, i.e. anti-corrosion and clogging issues regarding the pipeline, economic and environmental attributes are taken into consideration to increase the sustainability of the material selection. Four common commercial materials, including PVC, HDPE, carbon steel, and galvanized steel are considered as the available alternatives. The results of both IVTIFN-TOPSIS(E) and the IVTIFN-TOPSIS(M) models show that plastic pipes are better than the metal material alternatives. Especially, the latter indicates that HDPE is the optimal material, while PVC is near optimal for the pipeline design.

4.5 Grey Target Decision Making for Pipe Materials Selection

Landfill leachate is characterized by a wide range of pollutants and complex components, which are collected and transported by piping system to in-situ wastewater treatment facilities for advanced treatment [56, 128]. During the transport, the combined actions of physics, chemistry and microorganisms may give rise to pipe corrosion and scaling, and even cause bursting in severe cases, posing a risk to public health [95, 99, 74]. Thus, careful selection of materials is critical in the design of piping system of leachate collection and transport, thus to ensure safety management of landfill.

This study is to build an indicator system in the context of sustainable design that not only considers the functions of the material, but also take into account

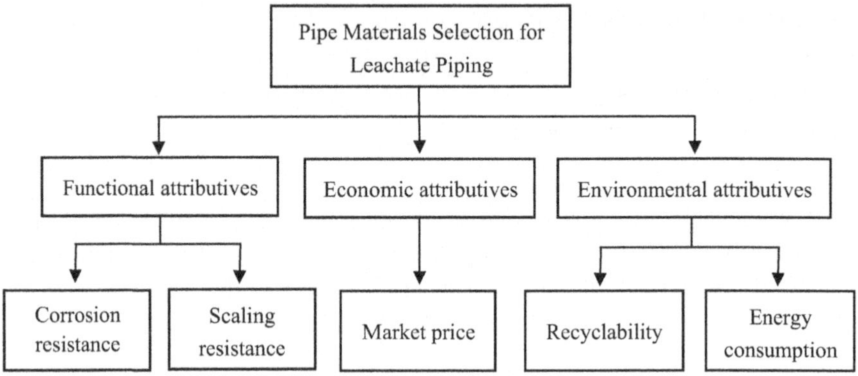

Fig. 4.8 Indicator system for pipe materials selection

its compatibility with economic and environmental performance, so as to perform precise selection of common commercial pipe materials by using multi-attribute decision making (MADM). An illustrative case example is given to demonstrate the model application and to help improve the experience-based material selection, thus enhancing the robustness of the design. It is also expected that this study can provide insight into material selection for leachate collection and transport, ultimately improving landfill management practice.

4.5.1 Criteria for Materials Selection

The hierarchy of an indicator system for pipe material selection is shown in Fig. 4.8. The functional attributes of the pipe material mainly take into account its corrosion resistance and scaling resistance, where the former evaluates the performance of metallic materials and the latter focuses on that of polymeric materials. In this study, the corrosion resistance is measured using scanning electron microscope (SEM) morphological observations, which rate the alternative materials from 1 to4 in terms of their degree of corrosion. The relative scaling resistance is characterized by the degree of scaling (i.e. weight increase), after a sample of each alternative material is immersed in leachate for a specified period of time. The economic attributes of the materials mainly focus on their market price, and the environmental attributes are indicated by their recyclability, and the energy consumption during processing.

4.5.2 Grey Target Decision Making

Let P be an option set with n alternative materials, $P = \{P_1, P_2, \ldots P_n\}$. A is an indicator set with m evaluation indicators $A = \{A_1, A_2, \cdots A_n\}$. The attribute value of option P_i for indicator set A_j is $\left[\underline{x}_{ij}, \overline{x}_{ij}\right](i = 1,2, \ldots n; j = 1,2, \ldots m)$. Let the interval grey number $x^t_{ij}(\otimes) \in \left[\underline{x}^t_{ij}, \overline{x}^t_{ij}\right](i = 1,2, \ldots n; j = 1,2, \ldots m)$ denote the attribute evaluation value for the option P_i at the tth indicator, where $\underline{x}^t_{ij}$ and $\overline{x}^t_{ij}$ are its lower and upper limits, respectively. Thus, the decision matrix of the option set P is given by [68]:

$$X = (x^t_{ij}(\otimes))_{n\times m} = \begin{bmatrix} \left[\underline{x}^t_{11}, \overline{x}^t_{11}\right] & \left[\underline{x}^t_{12}, \overline{x}^t_{12}\right] & \cdots & \left[\underline{x}^t_{1m}, \overline{x}^t_{1m}\right] \\ \left[\underline{x}^t_{21}, \overline{x}^t_{21}\right] & \left[\underline{x}^t_{22}, \overline{x}^t_{22}\right] & \cdots & \left[\underline{x}^t_{2m}, \overline{x}^t_{2m}\right] \\ \vdots & \vdots & & \vdots \\ \left[\underline{x}^t_{n1}, \overline{x}^t_{n1}\right] & \left[\underline{x}^t_{n2}, \overline{x}^t_{n2}\right] & \cdots & \left[\underline{x}^t_{nm}, \overline{x}^t_{nm}\right] \end{bmatrix} \tag{4.36}$$

Since each indicator has different dimensions, grey scale transformation is applied to nondimensionalize the indicators. The Benefit- and cost-type of indicators are standardized by using the Eqs. (4.37) and (4.38), respectively [28].

$$\underline{q}^t_{ij} = \frac{\underline{x}^t_{ij} - \underline{q}^t}{\overline{q}^t - \underline{q}^t}, \overline{q}^t_{ij} = \frac{\overline{x}^t_{ij} - \underline{q}^t}{\overline{q}^t - \underline{q}^t} \tag{4.37}$$

$$\underline{q}^t_{ij} = \frac{\overline{q}^t - \overline{x}^t_{ij}}{\overline{q}^t - \underline{q}^t}, \overline{q}^t_{ij} = \frac{\overline{q}^t - \underline{x}^t_{ij}}{\overline{q}^t - \underline{q}^t} \tag{4.38}$$

where $\overline{q}^t = \max\limits_{1\le i\le n}\max\limits_{1\le j\le m}\left\{\overline{x}^t_{ij}\right\}$ $\overline{\mathrm{r}}^{\mathrm{kt}} = \max\limits_{1\le \mathrm{i}\le \mathrm{n}}\max\limits_{1\le \mathrm{j}\le \mathrm{m}}\left\{\overline{\mathrm{x}}^{\mathrm{kt}}_{\mathrm{ij}}\right\}$ $\overline{\mathrm{r}}^{\mathrm{kt}} = \max\limits_{1\le \mathrm{i}\le \mathrm{n}}\max\limits_{1\le \mathrm{j}\le \mathrm{m}}\left\{\overline{\mathrm{x}}^{\mathrm{kt}}_{\mathrm{ij}}\right\}; \underline{q}^t = \min\limits_{1\le i\le n}\min\limits_{1\le j\le m}\left\{\underline{x}^t_{ij}\right\}$ $\underline{\mathrm{r}}^{\mathrm{kt}} = \min\limits_{1\le \mathrm{i}\le \mathrm{n}}\min\limits_{1\le \mathrm{j}\le \mathrm{m}}\left\{\underline{\mathrm{x}}^{\mathrm{kt}}_{\mathrm{ij}}\right\}$ $\underline{\mathrm{r}}^{\mathrm{kt}} = \min\limits_{1\le \mathrm{i}\le \mathrm{n}}\min\limits_{1\le \mathrm{j}\le \mathrm{m}}\left\{\underline{\mathrm{x}}^{\mathrm{kt}}_{\mathrm{ij}}\right\}$.

Both of the above measures are satisfied by $q^t_{ij}(\otimes) \in [0,1]$, and the larger the $q^t_{ij}(\otimes)$, the more ideal the indicator. Thus, a standard matrix $Q^t = \left(q^t_{ij}(\otimes)\right)_{n\times m}(1 \le i \le n, 1 \le j \le m)$ can be constructed for the option set P.

Assume that the weight vector of t indicators for the material alternatives is $\omega_t = (\omega_1, \omega_2, \ldots \omega_t)$, $\sum_{j=1}^{t}\omega_{tj} = 1, 0 \le \omega_{tj} \le 1$. To reduce the influence resulting from arbitrary decision, the coefficient of variation method is used in this study to adjust the subjective weighting determined by the expert scoring. The coefficient of variation determines the size of the weights by calculating the coefficient of variation between the information contained in each attribute indicator, which can overcome the equalization of the weighting distribution, and it is given by Hogg et al. [47]:

$$V_j = \frac{\sigma_j}{\overline{q}_j}(j = 1,2, \cdots m) \tag{4.39}$$

where V_j is the coefficient of variation of the jth indicator; σ_s is the standard deviation of the jth indicator $\sqrt{\frac{1}{N}\sum_{n=1}^{N}(q_j - \overline{q}_j)^2}$; $\overline{q}_j$ is the mean value of the jth indicator.

If the evaluation value is an interval number, then the mean of the upper and lower limits is calculated.

The weight μ_j of each indicator is given by:

$$\mu_j = \frac{V_j}{\sum_{j=1}^{m} V_j} \tag{4.40}$$

The integrated weight ω_t is:

$$\omega_t = \frac{\mu_j \omega_j}{\sum_{j=1}^{m} \mu_j \omega_j} \tag{4.41}$$

Once the weight ω_t is introduced, the decision matrix of option set P can be expressed as follows [68]:

$$Q = (q_{ij}(\otimes))_{n\times m} = \begin{bmatrix} [\underline{q}_{11}, \overline{q}_{11}] & [\underline{q}_{12}, \overline{q}_{12}] & \cdots & [\underline{q}_{1m}, \overline{q}_{1m}] \\ [\underline{q}_{21}, \overline{q}_{21}] & [\underline{q}_{22}, \overline{q}_{22}] & \cdots & [\underline{q}_{2m}, \overline{q}_{2m}] \\ \vdots & \vdots & & \vdots \\ [\underline{q}_{n1}, \overline{q}_{n1}] & [\underline{q}_{n2}, \overline{q}_{n2}] & \cdots & [\underline{q}_{nm}, \overline{q}_{nm}] \end{bmatrix} \tag{4.42}$$

where q_{ij} is the performance measure of the ith alternative material under the jth indicator.

At this point, the maximum interval number corresponding to each indicator is the optimal performance of that indicator, and the minimum is the least performance.

If $q_+^t = max\left\{(\underline{q}_{ij} + \overline{q}_{ij})/2|(1 \le i \le n, 1 \le j \le m)\right\}$, the corresponding measurement is given as $\left[\underline{q}_{i_0 j}^+, \overline{q}_{i_0 j}^+\right]$,

If $q_-^t = min\left\{(\underline{q}_{ij} + \overline{q}_{ij})/2|(1 \le i \le n, 1 \le j \le m)\right\}$, the corresponding measurement is given as $\left[\underline{q}_{ij}^-, \overline{q}_{ij}^-\right]$.

Thus, the optimal performance vector of the grey target decision-making, i.e. the "positive target", can be given by [67]:

$$\begin{aligned} q_0^+ &= \left\{q_+^1(\otimes), q_+^2(\otimes), \ldots, q_+^m(\otimes)\right\} \\ &= \left\{\left[\underline{q}_{i_0 1}^+, \overline{q}_{i_0 1}^+\right], \left[\underline{q}_{i_0 2}^+, \overline{q}_{i_0 2}^+\right], \ldots, \left[\underline{q}_{i_0 m}^+, \overline{q}_{i_0 m}^+\right]\right\} \end{aligned} \tag{4.43}$$

The least performance vector of the grey target decision-making, i.e. the "negative target", can be given by [67]:

$$\begin{aligned} q_0^- &= \left\{q_-^1(\otimes), q_-^2(\otimes), \ldots, q_-^m(\otimes)\right\} \\ &= \left\{\left[\underline{q}_{i1}^-, \overline{q}_{i1}^-\right], \left[\underline{q}_{i2}^-, \overline{q}_{i2}^-\right], \ldots, \left[\underline{q}_{im}^-, \overline{q}_{im}^-\right]\right\} \end{aligned} \tag{4.44}$$

Let the distance between the alternative P_i and the positive and negative target be L_i^+ and L_i^-, respectively, and the distance between the positive and negative target be L^0, they can be expressed as follows:

$$L_i^+ = 2^{-1/2}\Big[\omega_t(\underline{q}_{i1}^t - \underline{q}_{i_0 1}^+)^2 + \omega_t(\bar{q}_{i1}^t - \bar{q}_{i_0 1}^+)^2 + \cdots + \omega_t(\underline{q}_{ij}^t - \underline{q}_{i_0 j}^+)^2 + \omega_t(\bar{q}_{ij}^t - \bar{q}_{i_0 j}^+)^2\Big]^{1/2} \tag{4.45}$$

$$L_i^- = 2^{-1/2}\Big[\omega_1(\underline{q}_{i1}^t - \underline{q}_{i1}^-)^2 + \omega_1(\bar{q}_{ij}^t - \bar{q}_{ij}^-)^2 + \cdots + \omega_t(\underline{q}_{ij}^t - \underline{q}_{ij}^-)^2 + \omega_t(\bar{q}_{ij}^t - \bar{q}_{ij}^-)^2\Big]^{1/2} \tag{4.46}$$

$$L^0 = 2^{-1/2}\Big[\omega_1(\underline{q}_{i_0 1}^+ - \underline{q}_{i1}^-)^2 + \omega_1(\bar{q}_{i_0 1}^+ - \bar{q}_{i1}^-)^2 + \cdots + \omega_t(\underline{q}_{i_0 j}^+ - \underline{q}_{ij}^-)^2 + \omega_t(\bar{q}_{i_0 j}^+ - \bar{q}_{ij}^-)^2\Big]^{1/2} \tag{4.47}$$

Assume that the distance from the point where the option P_i is located to the positive target L_i^+ that satisfies $L_i^+ \leq L^0$, and that to the negative target L_i^- satisfies $L_i^- \leq L^0$. Thus, a triangle is formed in the space between the option and the positive and negative targets. The projection of the positive target on the connection of the positive and negative targets is expressed as the integrated target distance L_i [66]:

$$L_i = L_i^+ cos\theta = \frac{(L_i^+)^2 + (L^0)^2 - (L_i^-)^2}{2L^0} \tag{4.48}$$

The smallest the integrated target distance L_i, the best the option.

4.5.3 An Illustrative Case

Four common commercial drainage pipe materials were selected as alternative materials for leachate transfer, including seamless steel pipe (TC), galvanized steel pipe (RC), HDPE pipe, and PVC pipe. These pipe materials were purchased from a pipe supplier in Chengdu City. Their type specifications and compositions are shown in Table 4.16.

In order to measure the corrosion and scaling resistance of these pipe materials to the landfill leachate, the study has designed an immersion experiment, where samples of the above four pipe materials are immersed in the leachate sampled from a municipal landfill in Chengdu City. Following which, SEM and micro precision electronic balance are used to observe and weigh the samples periodically. The main components of the landfill leachate used in the immersion experiment are given in Table 4.17.

Table 4.16 Parameters of the pipe materials

Material	Outer diameter (mm)	Composition
TC	48.3	Fe(98%), C(0.17%), Si(0.25%), Mn(0.50%), Cr(0.25%), Ni(0.30%), Cu(0.25%)
RC	48.3	Fe(98%), Zn(0.65%),C(0.17%), Si(0.30%), P(0.02%), Mn(0.40%), Cu(0.25%)
HDPE	50	HDPE resin (89%), carbon black (2%), $CaCO_3$(3%), polyolefin elastomer (4%), antioxygen (0.4%), pe-wax (0.6%), calcium stearate (0.7%)
PVC	50	PVC resin (89%), $CaCO_3$(2.5%), lead salt stabilizer (3.5%), metal soaps (1.5%), pe-wax (0.6%), polyolefin impact modifier (2.5%), stearate (0.2%)

Table 4.17 Main components of the experimental landfill leachate

pH	COD (mg/L)	NH_4–N (mg/L)	DO (mg/L)	Ca^{2+} (mg/L)	Mg^{2+} (mg/L)	Cl^- (mg/L)	SO2- 4 (mg/L)
8.61	7589	738	6	2290	967	995	427

The specific experimental protocol is as follows: Samples labeled as #1 (TC), #2 (RC), #3 (HDPE), and #4 (PVC) with the same inner wall area were placed in a beaker containing 100 mL of leachate, respectively. Another comparable set of samples was immersed in 100 mL of distilled water. The whole experiment lasted for a total of 7 cycles, with 14 days per cycle. Four samples were taken for each type of pipe material, including triplicates for the leachate experiment and one in distilled water as a control. This resulted in a total of 112 samples. At the end of each experimental cycle, the samples for that cycle were removed, lyophilized under vacuum, weighed, and analyzed by SEM, whilst the samples from the other cycles remained immersing. The samples were weighed using an electronic balance (JA5003), and their surface morphologies were observed using SEM (JSM-7500F).

Table 4.18 shows the weight changes of the sample materials. The weight change of the TC sample was the largest, while that of the HDPE sample was the smallest. The ranking of weight increase for the four samples was: $\Delta S_{TC} > \Delta S_{RC} > \Delta S_{PVC} > \Delta S_{HDPE}$. The smallest weight increase indicates the best scaling resistance, scored by 4 points, while the largest weight increase indicates the least scaling resistance, scored by 1 point.

Figure 4.9 shows the surface morphology of different material samples immersed in the leachate. After the immersion for 98 days, the surface of TC became flocculent due to corrosion, and showed obvious corrosion pits, see Fig. 4.9a. The surface of RC became granular under the corrosive action of the leachate, and cracks had appeared, see Fig. 4.9b. Figure 4.9c shows that there was no obvious scaling or corrosion on the surface of HDPE. Figure 4.9d shows that the leachate corroded the PVC pipe material along the micropores on its surface, resulting in the continuous enlargement of these pores. The degree of corrosion on the four materials, in descending order,

Table 4.18 Weight of alternative material samples immersed in the leachate

T/ΔS	TC (g)	RC (g)	HDPE (g)	PVC (g)
T_1	0.015	0.016	0.009	0.009
T_2	0.018	0.010	0.013	0.026
T_3	0.028	0.023	0.001	0.006
T_4	0.020	0.020	0.005	0.005
T_5	0.001	0.004	0.004	0.005
T_6	0.014	0.006	0.007	0.001
T_7	0.003	0.003	0.001	0.000
ΔS_T	0.099	0.082	0.030	0.052

Note T indicates cycle; one cycle is 14 days
ΔS indicates the amount of scaling on the materials

was as follows: TC sample > RC sample > PVC sample > HDPE sample. Hence, the score of corrosion resistance was given as follows: HDPE 4; PVC 3; RC 2; TC 1.

The market prices of the pipe materials in this study were derived from the retail prices at which the experimental pipe materials were purchased. The data for energy

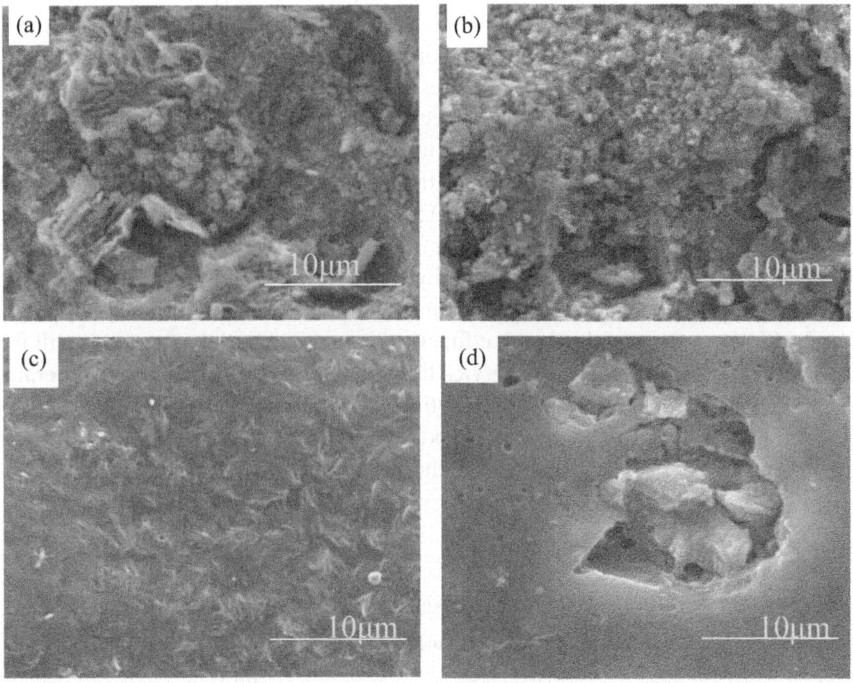

Fig. 4.9 SEM images of the material samples immersed in the leachate: **a** TC, **b** RC, **c** HDPE and **d** PVC

Table 4.19 Indicator data for the four pipe materials

Indicators	Alternative Material			
	TC	RC	HDPE	PVC
Corrosion resistance	0	0.333	1	0.667
Scaling resistance	0	0.246	1	0.681
Market price	[0.643, 0.785]	[0.285, 0.614]	[0, 0.285]	[0.785, 1]
Recyclability	[0.788, 1]	[0.623, 0.788]	[0.152, 0.188]	[0, 0.012]
Energy consumption	[0.837, 1]	[0.567, 0.875]	[0, 0.291]	[0.145, 0.378]

consumption and recyclability were derived from previous studies on the environmental impact of the alternative materials [8, 64, 76, 107]. Table 4.19 shows the normalized data corresponding to the different indicators.

4.5.4 Results

Five experienced engineers working in the investigated landfill were requested to give their preferences for the involved indicators. Their subjective judgment was adjusted using the coefficient of variation to obtain the weights for each indicator, as shown in the Table 4.20. The weights for functional, economic and environmental attributes were 0.488, 0.137 and 0.375, respectively.

The priority ranking was performed according to the target distance in ascending order, where the choice corresponding to the smallest target distance is considered as the optimal material. It is clear from Table 4.21 that the integrated target distances of the four pipe materials were 0.561, 0.504, 0.338 and 0.444, respectively; i.e. $L_{TC}>L_{RC}>L_{PVC}>L_{HDPE}$. This further implies that among the four alternative materials, HDPE was the best, followed by PVC, RC and the least was TC.

Among the attributes, the weight of functional indicators accounted for approximately 48.8%, reflecting the fact that function-oriented material selection is still the key to engineering design. As the service life of domestic waste landfills is generally maintained at more than 10 years, the lifetime of the pipe system for leachate collection and transport should be as long as possible [79]. The corrosion resistance and scaling resistance of pipe materials are the key factors affecting the service life of

Table 4.20 Weightings for each indicator

Indicators	Weighting value
Corrosion resistance	0.212
Scaling resistance	0.276
Market price	0.137
Recyclability	0.213
Energy consumption	0.162

Table 4.21 Target distance for different alternative materials

Alternative Materials	L^+	L^-	L
TC	0.701	0.540	0.561
RC	0.541	0.443	0.504
HDPE	0.537	0.700	0.338
PVC	0.540	0.550	0.444

pipelines. Among the four materials, the overall performance of HDPE was the best, while that of carbon steel pipe was the least. The findings are consistent with China's "Technical Code for Municipal Solid Waste Sanitary Landfill", which recommends the use of HDPE pipe materials for leachate piping system, thus confirming the rationality of the selection results. Mota et al. [80] compared the corrosion resistance of PVC, HDPE, stainless steel and copper for pipelines used in the food industry, and found that HDPE had the best performance. However, from the application of heating pipes, Zhao et al. [126] identified that PVC was the optimal material among three alternative materials, which differs from the findings of the study. This is primarily due to the different engineering application scenarios, which resulted in discrimination in the selection of evaluation indicators.

In addition, if only the order of the positive target distance is considered by using the conventional grey target decision-making, the selection result is: $L^+_{TC} > L^+_{RC} \approx L^+_{PVC} > L^+_{HDPE}$. In such case, the priority ranking of RC and PVC cannot be discriminated. This study improves the conventional grey target decision making based on spatial geometric correction, by taking integrated target distance into account, to obtain clear distinctions in the ranking result, which demonstrates its adaptability when applied to material selection.

4.5.5 Summary

Landfill leachate, due to its quantity and inherent risk, is generally collected and transported by a piping system for advanced treatment. During the piping, the pipe materials can react with leachate, resulting in corrosion and scaling. In order to reduce failures and mitigate possible consequences, this study provides an indicator system for material selection to aid in the design of the pipe system. This study contributes to a precise material selection system that can overcome the subjectivity of empirical material selection. Moreover, the grey target decision-making based on the integrated target distance can be served as an effective computational tool to discriminate performance of alternative materials in terms of their priority ranking, which can improve the accuracy of materials selection.

An indicator system is constructed based on the target criteria of functional, economic and environmental requirements, to perform the selection of pipe materials for landfill leachate transferring by grey target decision-making. Four common drainage pipe materials, namely, HDPE, PVC, galvanized steel and seamless steel were included in the selection. The result showed that HDPE was the best, followed by PVC, galvanized steel, and seamless steel, which confirmed the recommendation in China's technical code for landfill design. This further demonstrates the reasonableness of the model application, which can help in the design of pipe system for leachate collection and transport.

References

1. Akadiri PO, Olomolaiye PO, Chinyio EA (2013) Multi-criteria evaluation model for the selection of sustainable materials for building projects. Autom Constr 30:113–125
2. Alemi-Ardakani M, Milani AS, Yannacopoulos S, Shokouhi G (2016) On the effect of subjective, objective and combinative weighting in multiple criteria decision-making: a case study on impact optimization of composites. Expert Syst Appl 46:426–438
3. Allen P (2020) SATRA spotlight-colour fastness and rub testing. http://www.satra.co.uk/bulletin/article_view.php?id=740
4. Amah E (2012) Corporate culture and organizational effectiveness: a study of the nigerian banking industry. Eur J Bus Manage 4:212–229
5. Aoe T (2007) Eco-efficiency and eco design in electrical and electronic products. J Clean Prod 15:1406–1414
6. Arce ME, Saavedra Á, Míguez JL, Granada E (2015) The use of grey-based methods in multi-criteria decision analysis for the evaluation of sustainable energy systems: a review. Renew Sustain Energy Rev 47:924–932
7. Ashby MF (2005) Materials selection in mechanical design, 3rd edn. Butterworth-Heinemann, Oxford
8. Ashby MF (2009) Materials and the environment: eco-informed material choice. Elsevier, Oxford
9. Ashby MF, Johnson K (2013) Materials and design: the art and science of material selection in product design. Butterworth-Heinemann, Oxford
10. Ashby MF, Shercliff H, Cebon D (2013) Materials: engineering, science, processing and design. Butterworth-Heinemann, Oxford
11. Ashby MF, Shercliff H, Cebon D (2018) Materials: engineering, science, processing and design. Butterworth-Heinemann, Oxford
12. Athawale VM, Chakraborty S (2012) Material selection using multi-criteria decision-making methods: a comparative study. Proc Inst Mech Eng Part L J Mater Des Appl 226:266–285
13. Badurdeen F, Aydin R, Brown A (2018) A multiple lifecycle-based approach to sustainable product configuration design. J Clean Prod 200:756–769
14. Beng LG, Omar B (2014) Integrating axiomatic design principles into sustainable product development. Int J Precis Eng Manuf Green Technol 1:107–117
15. Bevilacqua M, Ciarapica FE, Giacchetta G (2011) Eco-design: a promising approach to sustainable production and consumption. Springer, London
16. Bharati SK, Singh SR (2018) Transportation problem under interval-valued intuitionistic fuzzy environment. Int J Fuzzy Syst 20:1511–1522
17. Binhussain MA, Marangoni M, Bernardo E, Colombo P (2014) Sintered and glazed glass-ceramics from natural and waste raw materials. Ceram Int 40:3543–3551
18. Bjørn A, Hauschild MZ (2013) Absolute versus relative environmental sustainability. J Ind Ecol 17:321–332
19. Bobyr MV, Milostnaya NA, Kulabuhov SA (2017) A method of defuzzification based on the approach of areas' ratio. Appl Soft Comput 59:19–32
20. Bovea MD, Vidal R (2004) Materials selection for sustainable product design: a case study of wood based furniture eco-design. Mater Des 25:111–116

21. Brezet H, Van Hemel C (1997) Eco-design: a promising approach to sustainable production and consumption. UNEP and TU Delft, Delft
22. Caliskan H, Kursuncu B, Kurbanoglu C, Guven SY (2013) Material selection for the tool holder working under hard milling conditions using different multi criteria decision making methods. Mater Des 45:473–479
23. Chan JWK, Tong TKL (2007) Multi-criteria material selections and end-of-life product strategy: grey relational analysis approach. Mater Des 28:1539–1546
24. Chang CH, Lin JJ, Lin JH, Chiang MC (2010) Domestic open-end equity mutual fund performance evaluation using extended TOPSIS method with different distance approaches. Expert Syst Appl 37:4642–4649
25. Chauhan A, Vaish R (2012) Magnetic material selection using multiple attribute decision-making approach. Mater Des 36:1–5
26. Clark G, Kosoris J, Hong LN, Crul M (2009) Design for sustainability: current trends in sustainable product design and development. Sustainability 1:409–424
27. Davis P, Sullivan E, Marlow D, Marney D (2013) A selection framework for infrastructure condition monitoring technologies in water and wastewater networks. Expert Syst Appl 40:1947–1958
28. Deng J (2010) Grey entropy and grey target decision making. J Grey Syst 22(1):1–4
29. Deng JL (1989) Introduction of grey system. J Grey Syst 1:1–24
30. Deng JL (1992) The essential methods of grey systems. Huazhong University of Science and Technology Press, Wuhan, China (in Chinese)
31. Deutz P, Neighbour G, McGuire M (2010) Integrating sustainable waste management into product design: sustainability as a functional requirement. Sustain Dev 18:229–239
32. Dornfeld DA (2014) Moving towards green and sustainable manufacturing. Int J Precis Eng Manuf Green Technol 1:63–66
33. Du D, Pang QH, Wu Y (2008) A modern comprehensive evaluation method and case selection. Qinghua University Press, Beijing, pp 115–116 (in Chinese)
34. Dubey R, Gunasekaran A, Ali SS (2015) Exploring the relationship between leadership, operational practices, institutional pressures and environmental performance: a framework for green supply chain. Int J Prod Econ 160:120–132
35. Dubois D (2011) The role of fuzzy sets in decision sciences: old techniques and new directions. Fuzzy Sets Syst 184:3–28
36. Edwards KL (1994) Towards more effective decision support in materials and design engineering. Mater Des 15(52):251–258
37. Ermolaeva NS, Castro MBG, Kandachar PV (2004) Materials selection for an automotive structure by integrating structural optimization with environmental impact assessment. Mater Des 25:689–698
38. Farag MM, Magd EEI (1992) An integrated approach to product design, materials selection and cost estimation. Mater Des 13(6):323–327
39. Figge F, Hahn T (2004) Sustainable value added—measuring corporate contributions to sustainability beyond eco-efficiency. Ecol Econ 48:173–187
40. Fiksel J (2009) Design for environment: a guide to sustainable product development, 2nd edn. McGraw-Hill Companies, New York, pp 83–115
41. Fong JP, Lee FJ, Lu IS, Uang SN, Lee CC (2014) Estimating the contribution of inhalation exposure to di-2-ethylhexyl phthalate (DEHP) for PVC production workers, using personal air sampling and urinary metabolite monitoring. Int J Hyg Environ Health 217:102–109
42. Fuente ADL, Pons O, Josa A, Aguado A (2016) Multi-criteria decision making in the sustainability assessment of sewerage pipe systems. J Clean Prod 112:4762–4770
43. Girubha RJ, Vinodh S (2012) Application of Fuzzy VIKOR and environmental impact analysis for materials selection of an automobile component. Mater Des 37:478–486
44. Giudice F, Rosa GL, Risitano A (2005) Materials selection in the life-cycle design process: a method to integrate mechanical and environmental performances in optimal choice. Mater Des 26:9–20

45. Gummesson E (2014) Commentary on "The role of innovation in driving the economy: lessons from the global financial crisis." J Bus Res 67:2743–2750
46. He B, Huang S, Wang J (2015) Product low-carbon design using dynamic programming algorithm. Int J Precis Eng Manuf Green Technol 2:37–42
47. Hogg RV, McKean JW, Craig AT (2012) Introduction to mathematical statistics, 7th edn. Pearson, New York
48. Huang HH, Zhang L, Liu ZF, Sutherland JW (2011) Multi-criteria decision making and uncertainty analysis for materials selection in environmentally conscious design. Int J Adv Manuf Technol 52:421–432
49. Hurst K (1992) Engineering design principles. Wiley, New York, pp 54–56
50. Ilg P, Hoehne C, Guenther E (2016) High-performance materials in infrastructure: a review of applied life cycle costing and its drivers—the case of fiber-reinforced composites. J Clean Prod 112:926–945
51. Jahan A, Barhraminasab M, Edwards KL (2012) A target-based normalization technique for materials selection. Mater Des 35:647–654
52. Jahan A, Edwards KL (2013) Multi-criteria decision analysis for supporting the selection of engineering materials in product design. Butterworth-Heinemann, Oxford
53. Jahan A, Mustapha F, Ismail MY, Sapuan SM, Bahraminasab M (2011) A comprehensive VIKOR method for materials selection. Mater Des 32:1215–1221
54. Jiang GM, Sun J, Sharma KR, Yuan ZG (2015) Corrosion and odor management in sewer systems. Curr Opin Biotechnol 33:192–197
55. Kalpakjian S, Schmid SR (2006) Manufacturing engineering and technology, 5th edn. Prentice Hall, Singapore
56. Kamaruddin MA, Yusoff MS, Rui LM, Isa AM, Zawawi MH, Alrozi R (2017) An overview of municipal solid waste management and landfill leachate treatment: Malaysia and Asian perspectives. Environ Sci Pollut Res 24:26988–27020
57. Kane G (2010) The three secrets of green business: unlocking competitive advantage in a low carbon economy. Earthscan, London
58. Khan ZA, Kamaruddin S, Siddiquee AN (2010) Feasibility study of use of recycled high density polyethylene and multi response optimization of injection moulding parameters using combined grey relational and principal component analyses. Mater Des 1:2925–2931
59. Khare S, Dell'Amico M, Knight C, McGarry S (2012) Selection of materials for high temperature latent heat energy storage. Sol Energy Mater Sol Cells 107:20–27
60. Kumar R, Jagadish, Ray A (2014) Selection of material for optimal design using multi-criteria decision making. Proc Mater Sci 6:590–596
61. Kwong CK, Jiang H, Luo XG (2016) AI-based methodology of integrating affective design, engineering, and marketing for defining design specifications of new products. Eng Appl Artif Intell 47:49–60
62. Law M, Bowie G (2007) Prediction of failure strain and burst pressure in high yield-to-tensile strength ratio linepipe. Int J Press Vessels Pip 84:487–492
63. Lee WS, Lin YC (2011) Evaluating and ranking energy performance of office buildings using grey relational analysis. Energy 36:2551–2556
64. Li Y (2014) Study on scaling and corrosion of HDPE pipeline material in leachate transportation. Master Thesis, Southwest Jiaotong University, Chengdu (in Chinese with English abstract)
65. Liao TW (2015) Two interval type 2 fuzzy TOPSIS material selection methods. Mater Des 88:1088–1099
66. Liu S, Yang Y, Forrest J (2017) Grey data analysis—methods, models and applications. Springer, Singapore
67. Liu SF, Lin Y (2010) Grey systems: theory and application. springer, London
68. Liu SF, Lin Y (2006) Grey information: theory and practical application. Springer, London
69. Ljungberg LY, Edwards KL (2003) Design, materials selection and marketing of successful products. Mater Des 24(7):519–529

70. Ljungberg LY (2007) Materials selection and design for development of sustainable products. Mater Des 28:466–479
71. Lo TP, Guo SJ (2010) Effective weighting model based on the maximum deviation with uncertain information. Expert Syst Appl 37:8445–8449
72. Madanhire I, Mugwindiri K, Mbohwa C (2015) Enhancing cleaner production application in fertilizer manufacturing: case study. Clean Technol Environ Policy 17:667–679
73. Maity SR, Chakraborty S (2013) Grinding wheel abrasive material selection using fuzzy TOPSIS method. Mater Manuf Processes 28:408–417
74. Makris KC, Andra SS, Botsaris G (2014) Pipe scales and biofilms in drinking-water distribution systems: undermining finished water quality. Crit Rev Environ Sci Technol 44(13):1477–1523
75. Mansor MR, Sapuan SM, Zainudin ES, Nuraini AA, Hambali A (2013) Hybrid natural and glass fibers reinforced polymer composites material selection using analytical hierarchy process for automotive brake lever design. Mater Des 51:484–492
76. Mao Y (2013) Based on multiple attribute decision optimization method of leachate transport pipeline material selection research. Master Thesis, Southwest Jiaotong University, Chengdu (in Chinese with English abstract)
77. McLennan JF (2004) The philosophy of sustainable design. Ecotone LLC, Kansas City, MO
78. Mezher T, Ajam M (2006) Integrating quality, environmental and supply chain management systems into the learning organisation. In: Sarkis (ed) Greening the supply chain. Springer, London, pp 67–85
79. Ministry of Construction of the People's Republic of China. (2009). Technical code for municipal solid waste sanitary landfill (CJJ 17-2004). China Architecture Publishing & Media Co., Ltd, Beijing, China (in Chinese)
80. Mota JM, Pereira A, Afonso MD (2018) Selection of materials for biofouling detection in cooling water systems. Water Sci Technol Water Supply 18:1162–1172
81. O'Connor TF, Zaretski AV, Savagatrup S, Printz AD, Wilkes CD, Diaz MI, Lipomi DJ (2016) Wearable organic solar cells with high cyclic bending stability: materials selection criteria. Sol Energy Mater Sol Cells 144:438–444
82. Pardo RJH, Bhamra T, Bhamra R (2012) Sustainable product service systems in small and medium enterprises (SMEs): opportunities in the leather manufacturing industry. Sustainability 4:175–192
83. Petit-Boix A, Sanjuan-Delmás D, Gasol CM, Villalba G, Suárez-Ojeda ME, Gabarrell X, Josa A, Rieradevall J (2014) Environmental assessment of sewer construction in small to medium sized cities using life cycle assessment. Water Resour Manag 28:979–997
84. Piccinno F, Hischier R, Saba A, Mitrano D, Seeger S, Som C (2016) Multi-perspective application selection: a method to identify sustainable applications for new materials using the example of cellulose nanofiber reinforced composites. J Clean Prod 112:1199–1210
85. Prasad K, Chakraborty S (2013) A quality function deployment-based model for material selection. Mater Des 49:525–535
86. Rahman S, Odeyinka H, Perera S, Bi Y (2012) Product-cost modelling approach for the development of a decision support system for optimal roofing material selection. Expert Syst Appl 39:6857–6871
87. Rao RV, Davim JP (2008) A decision-making framework model for material selection using a combined multiple attribute decision-making method. Int J Adv Manuf Technol 35:751–760
88. Rao RV, Patel BK (2010) A subjective and objective integrated multiple attribute decision-making method for material selection. Mater Des 31:4738–4747
89. Ray DL, Guzzo L (1993) Environmental overkill: whatever happened to common sense. Harper Collins, New York, NY
90. Reddy GP, Gupta N (2010) Material selection for microelectronic heat sinks: an application of the Ashby approach. Mater Des 31:113–117
91. Roscoe S, Cousins PD, Lamming RC (2016) Developing eco-innovations: a three-stage typology of supply networks. J Clean Prod 122:1948–1959

92. Saaty TL (2001) Analytic hierarchy process. In: Harris G (ed) Encyclopedia of operations research and management science. Springer, London, UK, pp 19–28
93. Şengül Ü, Eren M, Shiraz SE (2015) Fuzzy TOPSIS method for ranking renewable energy supply systems in Turkey. Renew Energy 75:617–625
94. Serafini M, Russo D, Rizzi C (2015) Multi criteria material selection for eco-design. Comput Aided Des Appl 12:526–536
95. Shaha BN, Meeroff DE, Kohn K, Townsend TG, Schert JD, Mayer N, Schultz R, Telson J (2019) Effect of electronic water treatment system on calcium carbonate scale formation in landfill leachate collection piping. J Environ Eng 145(9):04019052
96. Shanian A, Savadogo O (2006) A material selection model based on the concept of multiple attribute decision making. Mater Des 27:329–337
97. Smith P (2007) The relationship between management accounting, profitability and operations in an uncertain world: evidence from literature and practice. Ph.D. Thesis, Cranfield University, Cranfield, UK
98. Stewart WH Jr, Carland JC, Carland JW, Watson WE, Sweo R (2003) Entrepreneurial dispositions and goal orientations: a comparative exploration of United States and Russian entrepreneurs. J Small Bus Manage 41:27–46
99. Stibinger J (2017) Approximation of clogging in a leachate collection system in municipal solid waste landfill in Osecna (Northern Bohemia, Czech Republic). Waste Manage 63:131–142
100. Tamani N, Mosse P, Croitoru M, Buche P, Guillard V, Guillaume C, Gontard N (2015) An argumentation system for eco-efficient packaging material selection. Comput Electron Agric 113:174–192
101. Testa F, Iraldo F (2010) Shadows and lights of GSCM (green supply chain management): determinants and effects of these practices based on a multinational study. J Clean Prod 18:953–962
102. Tjader Y, May JH, Shang J, Vargas LG, Gao N (2014) Firm-level outsourcing decision-making: a balanced scorecard-based analytic network process model. Int J Prod Econ 147:614–623
103. Toutanji H (2001) Dempsey, S. Stress modeling of pipelines strengthened with advanced composites materials. Thin Walled Struct 39:153–165
104. Tscoulfas TG, Pappis CP (2008) A model for supply chains environmental performance analysis and decision making. J Clean Prod 16:1647–1657
105. Tzeng GH, Huang JJ (2001) Multiple attribute decision making: methods and applications. CRC Press, Boca Raton, FL
106. Tzeng ML (2010) Using linguistic preferences and grey relational analysis to evaluate the environmental knowledge management capacity. Expert Syst Appl 37:70–81
107. Vahidi E, Jin E, Das M, Singh M, Zhao F (2016) Environmental life cycle analysis of pipe materials for sewer systems. Sustain Cities Soc 27:167–174
108. Van Kesteren IEH (2008) Product designers' information needs in materials selection. Mater Des 29:133–145
109. Vezzoli C, Manzini E (2008) Design for environmental sustainability. Springer, London
110. Wang HF, Gupta SM (2011) Green supply chain management: product life cycle approach. McGraw-Hill, New York, NY
111. Wang JQ, Wu JT, Wang J (2014) Interval-valued hesitant fuzzy linguistic sets and their applications in multi-criteria decision-making problems. Inf Sci 288:55–72
112. Wang ZW, Lei TZ, Chang X, Shi XG, Xiao J, Li ZF, He XF, Zhu JL, Yang SH (2015) Optimization of a biomass briquette fuel system based on grey relational analysis and analytic hierarchy process: a study using cornstalks in China. Appl Energy 157:523–532
113. Wang CN, Nguyen VT, Duong DH, Thai HTN (2018) A hybrid fuzzy analysis network process (FANP) and the technique for order of preference bysimilarity to ideal solution (TOPSIS) approaches for solid waste to energy plant location selection in Vietnam. Appl Sci 8:1100
114. Wei GW (2011) Grey relational analysis model for dynamic hybrid multiple attribute decision-making. Knowl Based Syst 24:672–679

115. McDonough W, Braungart M (2002) Cradle to cradle: remaking the way we make things. North Point Press, New York, NY
116. Winckle S (2020) SATRA Spotlight-colour transfer from handbags. http://www.satra.co.uk/spotlight/article_view.php?id=129
117. Winston WL, Albright SC (2008) Practical management science. South-Western Cengage Learning, Boston, MA
118. Xu G, Yang YP, Lu SY, Li L, Song X (2011) Comprehensive evaluation of coal-fired power plants based on grey relational analysis and analytic hierarchy process. Energy Policy 39:2343–2351
119. Yang SS, Nasr N, Ong SK, Nee AYC (2015) Designing automotive products for remanufacturing from material selection perspective. J Clean Prod 153(1):570–579
120. Zhai L, Khoo L, Zhong Z (2008) Design concept evaluation in product development using rough sets and grey relational analysis. Expert Syst Appl 3(36):7072–7079
121. Zhang ZX, Wang GX, Massarotto P, Rudolph V (2006) Optimization of pipeline transport for CO_2 sequestration. Energy Convers Manage 47:702–715
122. Zhao R (2012) Environmental risk management system design for hazardous waste materials. Ph.D. Thesis, University of Hull, Hull, UK
123. Zhao R, Neighbour G, Deutz P, McGuire M (2012) Materials selection for cleaner production: an environmental evaluation approach. Mater Des 37:429–434
124. Zhao R, Neighbour G, McGuire M, Deutz P (2013) A software based simulation for cleaner production: a game between manufacturers and government. J Loss Prev Process Ind 26:59–67
125. Zhao R, Peng D, Li Y (2015) An Interaction between government and manufacturer in implementation of cleaner production: a multi-stage game theoretical analysis. Int J Environ Res 9:1069–1078
126. Zhao R, Su H, Chen X (2016) Commercially available materials selection in sustainable design: an integrated multi-attribute decision making approach. Sustainability 8:79
127. Zhao R, Xi B, Liu Y (2017) Economic potential of leachate evaporation by using landfill gas: a system dynamics approach. Resour Conserv Recycl 124:74–84
128. Zhao R, Wang X, Chen X, Liu Y (2019) Impacts of different aged landfill leachate on PVC corrosion. Environ Sci Pollut Res 26(18):18256–18266
129. Anojkumar L, Ilangkumaran M, Sasirekha V (2014) Comparative analysis of MCDM methods for pipe material selection in sugar industry. Expert Syst Appl 41: 2964-2980.

Chapter 5
Application of MCDM to Solid Waste Management

5.1 Introduction

With the growth of population, municipal waste disposal has become a major issue in urbanization [50, 65]. Despite great efforts have been made in waste reduction by reuse or recycling, in most cases, waste materials treatment appears to be relatively straight forward and, in many cases, results in landfill disposal [55]. This is relatively simplistic since most, but not all, materials can be considered harmful or 'hazardous' to the environment to a greater or lesser extent. That is, most materials sent to landfill have an environmental impact.

Sustainable consumption and production calls for urban transition in response to the crisis of climate change [123, 83]. Currently, the greenhouse gas emissions (GHGs) from municipal waste disposal accounts for 3% of global emissions [20]. As countries strive for economic development and urbanization, they are also faced with the urgent issues of designing proper municipal solid waste (MSW) disposal plans and identifying potential ways to reduce emissions [29]. Landfilling, incineration, and composting are currently the most common ways of MSW disposal [14, 34, 125]. As conditions vary from country to country, there is a dilemma occurs in the choice of disposal ways [33]. For example, the investment and operational costs of incineration and anaerobic digestion are generally higher than those of landfilling, but the latter contributes abundant landfill gases (LFGs and leachate, which makes it difficult with respect to the reduction of GHGs and may even cause secondary environmental pollution [3, 7, 118]. Therefore, it is essential to prioritize the coordination between the economy, environment, and society when formulating a municipal solid waste management plan to increase its sustainability.

Legislation is a requisite for guiding the municipal solid waste management. For example, the " Law of the People's Republic of China on the Prevention and Control of Environmental Pollution Caused by Solid Waste" is the mail legislation in China on solid waste management [56]. It stipulates that local governments should

R. Zhao, *Multi-criteria Decision-Making Approaches to Sustainable Consumption and Production*, Urban Sustainability, https://doi.org/10.1007/978-981-96-2775-2_5

formulate and promulgate specific measures for waste supervision and administration, including the establishment of collection, transportation, and treatment facilities [114, 115]. This ensures the importance of analyzing the possible status quo of waste disposal to build appropriate planning scenarios for future management practices.

This Chapter aims to develop sustainable waste disposal by minimizing environmental risk and the associated carbon emissions, as well as maximizing public acceptance, through the application of multi-criteria decision making, to help the policy makers identify novel solutions and mitigation strategies. In this context, Sect. 5.2 introduces a conceptual planning scenario for landfill gas and leachate control by using the landfill gas as the combustion promoter for incineration of the leachate, to minimize the potential environmental risk posed by landfill disposal. On this basis, Sect. 5.3 further models the economic potential of a regional incineration plant using landfill gas to evaporate waste leachate, to investigate whether such a mode can be incorporated into future management practice. Section 5.4 models the interactions among stakeholders with conflicts regarding the operation of a waste incineration plant, including the operational enterprise of incineration plant, the local government and nearby residents, to identify the appropriate strategic action for alleviating the conflicts. Section 5.5 takes the sudden environmental pollution accidents caused by the hazardous materials railway transportation into account, to present a design for an environmental risk information management system, to support safe and reliable logistics.

5.1.1 A Plan for Sustainable Solid Waste Disposal

The main environmental risks associated with the landfill disposal are the production of leachate and landfill gas. Depending on the type and age of the landfill, leachate is generated at different rates. Leachate can be described as the liquid that drains from a landfill in the form of suspended solids, dissolved components of waste materials, byproducts from degradation etc. through percolation within the landfill [64, 98]. Because it can contain harmful and complex compounds i.e. organic matter leading to high chemical oxygen demand (COD) and biological oxygen demand (BOD), and heavy metals such as iron, copper, etc., leachate can penetrate through soils or groundwater, and thus adversely affect plant growth, contaminate the drinking water and impair the ecosystem [91, 98]. Moreover, leachate treatment is one of the most difficult problems to manage from the start of landfill operations until potentially many decades after closure [13]. Landfill gas, if uncontrolled, is considered to be one of the major sources of greenhouse gas emissions, especially in producing large amounts of methane, which is about 21 times more potent than CO_2 [17]. Thus, landfill should be used as a last resort and only for certain types and amounts of waste [17]. However, as long as landfill still remains a disposal option, leachate and gas will continue to be generated and thus remain a potential environmental risk to the public.

This section provides a conceptual alternative treatment system to control leachate and landfill gas, ultimately reducing the environmental risk and saving energy and disposal cost. It is expected that this study can provide an insight for landfill operators and engineers to promote the transformation from the conceptual work to the real achievement.

5.1.2 *Design for Landfill Gas and Leachate Control*

The principal of "using waste to treat waste" is proposed to aid in the conceptual design of landfill leachate control byconsidering one type of waste material that can be recycled or reused as a "feedstock", that can be used as a direct or indirect source to treat other kinds of waste. Here, it is suggested that certain by-products from landfill disposal can be used in this way, in this case using landfill gas to treat leachate, noting the successful experience of leachate evaporation directly fueled by landfill gas. However, this proposed conceptual prototype chooses to use incineration instead of evaporation technology, and focuses on exploring a possible relationship between the existing landfill and an incineration plant located nearby, and comparing this to a conventional leachate evaporation system located at the landfill. In particular, landfill gas is considered as an indirect use by producing methane to improve incinerator combustion in comparison with direct use for evaporation. Although the use of landfill gas for leachate thermal treatment is indicated by the reduction of fuel and operations cost, the detailed economic feasibility has not been reported in the current literature. This study will provide insight to analyze the potential economic benefits derived from the proposed conceptual prototype.

In Fig. 5.1, three typical waste disposal approaches are invoked: recycling or reuse, incineration and sanitary landfill. Whilst, in order to relieve the pressure on sanitary landfill, recycling or reuse is the preferred option in terms of the "hierarchy of waste management", waste components with significant calorific value, together with hazardous waste, can be sent directly to the incineration plant for thermal treatment. Through the energy recovery system, landfill gas can be purified as methane and collected as a combustion support for the incinerators, helping to reduce the requirement for combustion improvers such as diesel, gasoline, etc. In an "energy from waste" scheme, the incineration plant can be used not only to generate electricity, but also to thermally treat the landfill leachate.

5.1.3 *Analysis for the Plan Feasibility*

Suppose that municipal waste is treated in a unified planning area, in which a landfill and an incineration plant are located as a waste treatment 'park'. Thus, the total economic benefit is the sum of the contributions from the landfill and the incineration plant. Figure 5.2 illustrates a possible relationship between a landfill and

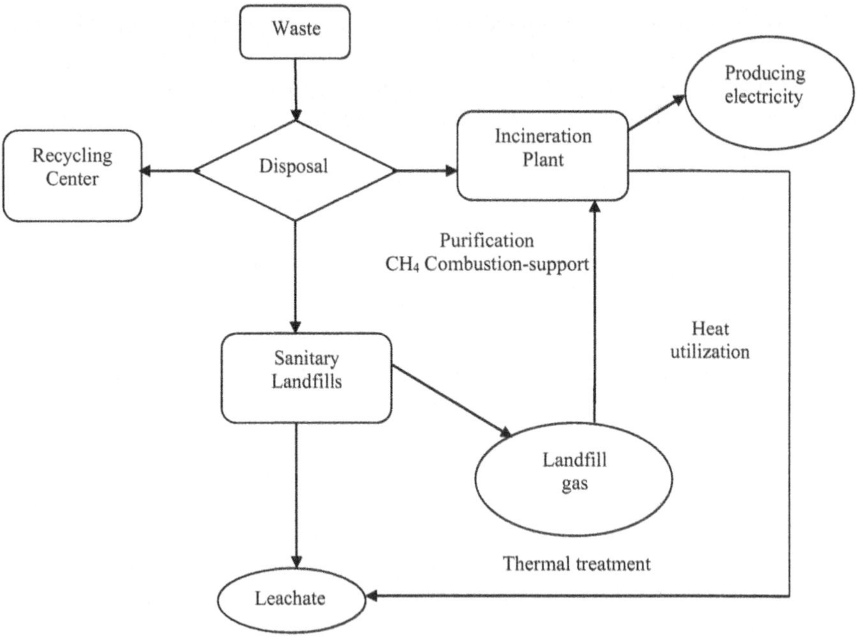

Fig. 5.1 Conceptual model for landfill gas and leachate control

an associated incineration plant. It is assumed that the landfill and the incineration plant operate independently in the initial stage. Subsequently, an energy recovery system is built to collect and purify the landfill gas for sale. The leachate is first pre-treated and then transferred to the municipal sewage plant for further treatment. In addition, diesel fuel is used as the combustion promoter to incinerate the waste and the energy is used to generate electricity.

In this study, the total economic benefit analysis is carried out by the comparing the original benefit assuming separate operation, and the expected benefit derived from our conceptual prototype. Let O denote the original total economic benefit of the proposed waste treatment park with no connection between landfill and incinerator, which can be expressed as follows.

$$O = G_L + G_I + P_{ele} + P_{\lg} - C_{L\cdot m} - C_{I\cdot m} - C_{leachate} - C_{fuel} \tag{5.1}$$

where

G_L annual waste gate fee for landfill
G_I annual waste gate fee for incineration
P_{ele} annual profit for electricity generation from incinerator
P_{lg} annual profit from selling prepared landfill gas
C_{Lm} annual maintenance cost for landfill
C_{Im} annual maintenance cost for incinerator

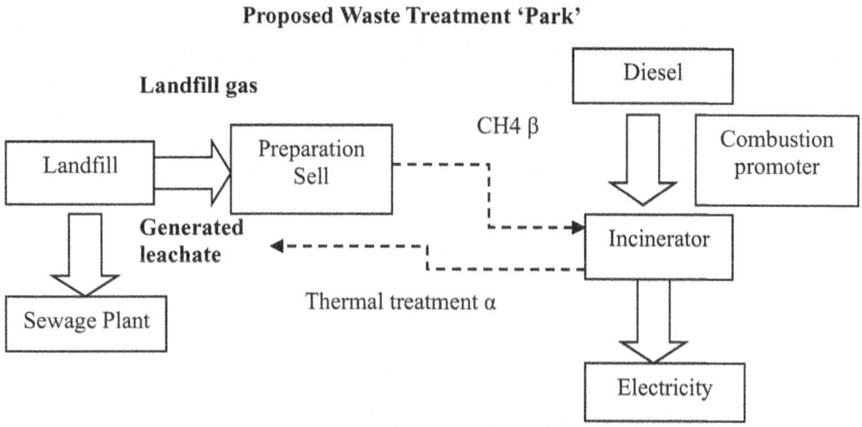

Fig. 5.2 Possible connection between landfill and incineration plant

$C_{leachate}$ annual cost paid to sewage plant for leachate treatment
C_{fuel} annual cost for diesel purchasing.

In the Eq. (5.9), suppose that the parameters such as G_L, G_I, P_{ele}, C_{Lm}, C_{Im}, will not change upon the implementation of the conceptual design. However, P_{lg}, $C_{leachate}$ and C_{fuel} will be changed accordingly, and can be obtained as follows.

$$P_{\lg} = M_{\lg} \cdot q \tag{5.2}$$

$$C_{leachate} = M_{leachate} \cdot C_{t \cdot l} \tag{5.3}$$

$$C_{fuel} = I_{diesel} \cdot P_{industrial} \tag{5.4}$$

where

M_{lg} annual output of purified landfill gas (CH_4) through energy recovery system
q selling price of purified landfill gas (CH_4), per tonne
$M_{leachate}$ annual output of landfill leachate
C_{tl} sewage treatment cost for leachate, per tonne
I_{diesel} annual diesel consumption for incineration
$P_{industrial}$ industrial price of diesel, per tonne

Substituting Eqs. (5.10), (5.11) and (5.12), Eq. (5.9) becomes:

$$O = G_L + G_I + P_{ele} + M_{\lg} \cdot q - C_{Lm} - C_{Im} - M_{leachate} \cdot C_{tl} - I_{diesel} \cdot P_{industrial} \tag{5.5}$$

With the proposed conceptual model to be implemented in the future, there is a possible connection between the landfill and the incinerator as reflected by the dashed arrows (see Fig. 5.2). A percentage β of the purified landfill gas will be sent to the incinerator as the combustion promoter, as well as a percentage α of the leachate consumed by thermal treatment. Let E denote the total expected cost benefit of the conceptual model, which can be derived as follows.

$$E = G_L + G_I + P_{ele} + (1-\beta)M_{\text{lg}} \cdot q - C_{Lm} - C_{Im} - (1-\alpha)M_{leachate} \cdot C_{tl} - \left(I_{diesel} - \frac{\beta M_{\text{lg}} J_1}{J_2}\right) \cdot P_{industrial} \tag{5.6}$$

where

J_1 complete combustion value of purified landfill gas (CH_4)
J_2 complete combustion value of diesel.

The comparison between the original and expected benefits can be obtained by the subtraction of Eqs. (5.14) and (5.13), as show below.

$$E - O = \alpha M_{leachate} \cdot C_{tl} + \beta M_{\text{lg}}\left(\frac{J_1}{J_2} \cdot P_{industrial} - q\right) \tag{5.7}$$

As the complete combustion value of CH_4 (55.5 MJ/kg) is higher than that of diesel (44.8 MJ/kg) $J_1 > J_2$, $J_1/J_2 \cdot P_{\text{industrial}} - q > 0$ since at least $P_{industrial} \geq q$. Thus, $E - O > 0$ which means that the expected economic benefit is higher than the initial one. Essentially, this analysis indicates that the conceptual prototype for waste disposal will bring a greater economic return.

This study mainly focuses on the situation of old saturated landfills that have been supplemented by an incineration facility in order to extend their life, with the hypothesis that LFG is a stable source of the combustion substitute for leachate incineration. But there are uncertainties in assessing the technical feasibility of the alternative system, as the production of landfill gas is a complex process that depends on many factors, such as the characteristics of wastes (e.g. biodegradable and organic, moisture content), the presence of micro-organisms, suitable aerobic and anaerobic conditions, the degree of initial compaction, etc. [80, 98]. The rate of LFG generation also changes over time, and may be influenced by the type of landfill (e.g. aerobic, semi-aerobic, anaerobic, or bioreactor landfill), waste density etc. [2, 80].

The developed countries are expected to reduce landfill disposal and thus landfill gas emissions, by improving waste recycling and reusing technologies [36]. However, when a landfill site enters the closure phase of its life, although the generation of LFG may decrease, it may still continue for decades after the closure [62]. This argument proves that landfill gas still could be a stable source for energy utilization, except for the phase of intensive gas production, which may not affect the operation of the proposed alternative system [99]. For example, empirical studies have shown that the energy from extracted waste from a landfill can still reach the level required for incineration after closure [30, 62].

As stated previously, leachate is a hazardous liquid waste containing non-biodegradable organic matter, ammonia nitrogen, heavy metals, chlorinated organic compounds and inorganic salts [41]. Thus, the long-term behaviour of leachate should be considered in the operation of the landfill in order to provide appropriate treatment until its quality meets environmental criteria. It is not recommended in the proposed treatment system to incinerate wastes with high chloride concentration to avoid the formation of dioxins. However, tremendous progress in new technologies for landfill leachate treatment has been reported in recent years. In the last two decades, landfills in developed countries have been equipped with highly efficient on-site landfill treatment systems, such as membrane bioreactor (MBR) and advanced oxidation technologies [69, 93]. These technologies have been shown to be a potential alternative for conventional and combined landfill leachate treatment due to their high removal efficiencies (for both BOD and COD), flexibility (for both old and young landfills), low residuals, etc. [1, 41, 69, 82], and could be considered as an auxiliary approach for the proposed system.

With regard to the economic analysis, the proposed alternative treatment system is based on the assumption that the energy recovery system required to purify and collect LFG has already been constructed at the landfill site. This requires the installation of landfill gas collection pipelines from the early stages of landfill construction and the construction of a methane gas refinery. Thus, the initial investment and further maintenance costs should be taken into account to validate the economic analysis. This highlights the need to consider this proposal in the context of an integrated 'waste treatment park', rather than as a subsequent development in conjunction with an existing landfill.

5.1.4 Summary

This study provides a conceptual framework that combines the technology of landfill with incineration for waste disposal, in which the recycled landfill gas is transferred to an incinerator as a combustion promoter and the heat is used to thermally treat the leachate, in order to reduce the energy consumption and disposal costs. According to the feasibility analysis, the proposed conceptual model can be designed based on existing technology, and could bring greater economic returns in the long term. With the development of technology for landfill gas reutilization and leachate incineration, the conceptual prototype can be realized in the near future and provide a template for leachate and landfill gas control.

5.2 Economic Potential of Leachate Evaporation by Using Landfill Gas

In order to minimize the risk posed by waste disposal, efforts should be focused on leachate treatment and landfill gas recovery. However, most of the landfill management practices treat these two issues separately, with little attention to their connections [118]. For instance, leachate, in some developing countries, is still piped to the local sewage plant for advanced treatment by piping system, due to its comparatively low cost [22]. But this alternative has an inefficient performance, not only increases the carrying capacities of municipal sewage treatment plant, but also causes complexities in sewage treatment, such as any heavy metals or non-biodegradable organics mixed [105]. In addition, most of the landfills prefer to treat LFGs by torching combustion rather than extraction for recovery [8, 72]. From the perspective of sustainable waste disposal, a combination of different treatments is recommended by the "Management of Waste Hierarchy", which aims not only to minimize waste volume and emissions, but also to maximize economic values related to resource recycling and energy recovery (El-Hagger 2007, [73]). In this context, an alternative treatment system is proposed by using LFG to evaporate waste leachate, thus promoting sustainable waste disposal. A system dynamics (SD) approach is used to model the economic potential of the alternative system.

This section mainly focuses on investigating the economic performance of a local incineration plant using LFG as a combustion improver to provide a template for its leachate control. A municipal incineration plant, located at Chengdu City, is considered to purchase the LFG as the combustion promoter recovered from a landfill nearby, and the residual heat is used to evaporate its waste leachate. A series of simulation scenarios are provided to examine the economic revenues affected by changing the leachate evaporation and LFG utilization, by which the optimal scenario is identified to be incorporated into the future management practice. It is expected that the study can provide insight into the development of an alternative treatment mode of waste leachate, thus promoting sustainable development.

5.2.1 Model Construction

An integrated treatment system is proposed in which the incineration plant purchases the purified LFG from its nearby landfill as an alternative combustion promoter, and treats the associated leachate by means of on-site surface spraying and evaporation. The SD model is built to seek the optimal economic performance of the incineration plant, while maximizing the ratio of LFG purchase α (the purchase amount of purified LFG accounted for the total purified LFG) and the ratio of waste leachate treatment by using evaporation β.

A number of key variables are considered, such as annual municipal waste received for incineration, purchased amount of purified LFG, annual leachate received for evaporation, annual power generation, calorific value of waste, etc., and their interactions are represented by a casual loop diagram. These variables not only determine the fundamental units of the proposed model, but also define the system boundary. The casual loop diagram is intended to help decision makers to understand the structure and the function of a system, which serves as a visualized conceptual model to represent the interrelationships among the significant variables [43, 86]. By using the Vensim software package, two typical clusters of feedback loops are defined, i.e. the reinforcing loop and balancing loop, as shown in Fig. 5.3. Each feedback loop is a closed loop in which one variable has either a positive or negative impact on another variable. There are two reinforcing loops and two balancing loops in the diagram.

The first reinforcing loop is that the power generation is increased by receiving more waste for incineration to improve the economic performance, and the more economic revenue received will lead to an increase in waste for incineration. The second reinforcing loop is that the increase in leachate evaporation saves the conventional treatment cost to improve the economic performance of the incineration plant, which gives rise to receiving more municipal waste for incineration, thus generating more waste leachate. With regard to the balancing loops, the first is that the calorific value of the waste decreases as the amount of leachate for incineration increases, which results in a decrease in the economic revenue. In this case, the waste received for incineration will be reduced in order to reduce the associated

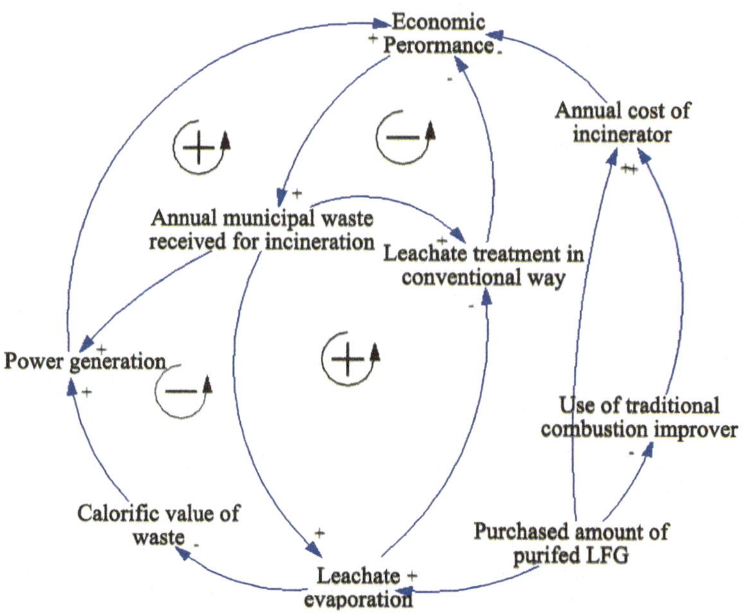

Fig. 5.3 Causal loop diagram of the combined system for leachate evaporation

leachate. The second is that the increase of waste for incineration may bring additional cost for leachate treatment in the conventional way, thus having a negative impact on the economic revenue of the incineration plant, by which the incineration capacity will be decreased.

Based on the analyses above, these two feedback loop clusters are defined as follows.

Reinforcing loop clusters: Annual municipal waste received for incineration → Power generation → Economic performance → Annual municipal waste received for incineration; Leachate evaporation → Leachate treatment in conventional way → Economic performance → Annual municipal waste received for incineration → Leachate evaporation.

Balancing loop clusters: Leachate evaporation → Calorific value of waste → Power generation → Economic performance → Annual municipal waste received for incineration → Leachate evaporation; Annual municipal waste received for incineration → Leachate treatment in conventional way → Economic performance → Annual municipal waste received for incineration.

A stock-flow diagram is constructed to perform mathematical simulation and quantitative analysis, thus to quantify the impacts of the identified variables [113]. Unlike the causal-loop diagram, a number of auxiliary variables are added to the stock-flow diagram to ensure that the above defined relationships can be transformed into appropriate mathematical equations [43]. Through the application of the Stella software package, a stock-flow diagram is shown in Fig. 5.4. In order to facilitate the understanding of the proposed system, the key variables in the model are described in Table 5.1.

The system is further divided into two sub-systems, including a sub-system of LFG generation and a sub-system of incinerator operation, indicated as two stocks.

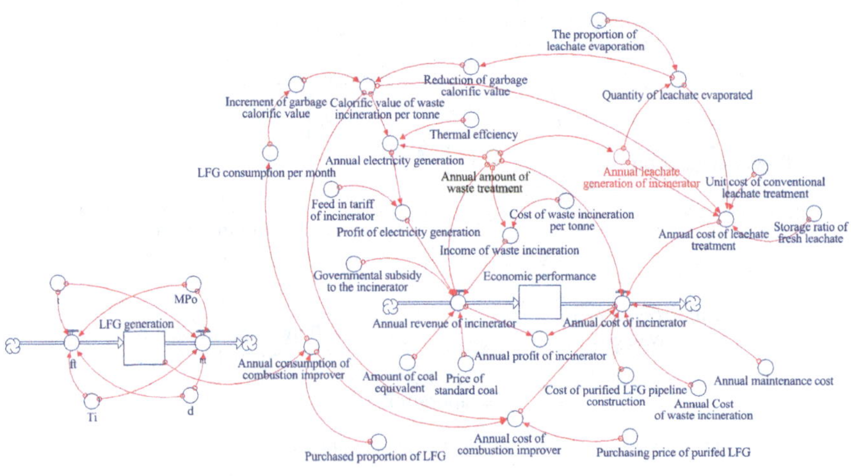

Fig. 5.4 Stock-flow diagram of the combined system

Table 5.1 Key variables of in the stock-flow diagram

Key variable	Variable type	Key variable	Variable type
LFG generation	Stock	Annual amount of waste treatment	Auxiliary variable
Ft	Flow	Annual leachate generation of incinerator	Auxiliary variable
Ftt	Flow	The proportion of leachate evaporation	Auxiliary variable
Ti	Constant	Quantity of leachate evaporated	Auxiliary variable
D	Constant	Reduction of garbage calorific value	Auxiliary variable
MP_0	Constant	Calorific value of waste incineration per tonne	Auxiliary variable
Annual consumption of combustion improver	Auxiliary variable	Annual cost of leachate treatment	Auxiliary variable
Purchased proportion of LFG	Auxiliary variable	Unit cost of conventional leachate treatment	Constant
Annual cost of combustion improver	Auxiliary variable	Storage ratio of fresh leachate	Constant
Feed in tariff	Constant	Annual electricity generation	Auxiliary variable
Purchasing price of purified LFG	Constant	Thermal efficiency	Constant
LFG consumption per month	Auxiliary variable	Profit of electricity generation	Auxiliary variable
Increment of garbage calorific value	Auxiliary variable	Income of waste incineration	Auxiliary variable
Price of waste incineration per tonne	Constant	Annual revenue of incinerator	Flow
Governmental subsidy to the incinerator	Constant	Amount of coal equivalent	Constant
Price of standard coal	Constant	Annual cost of incinerator	Flow
Annual maintenance cost	Constant	Cost of purified LFG pipeline construction	Constant
The proportion of leachate evaporation	Auxiliary variable	Economic performance of incineration plant	Stock

The investigated landfill is responsible for the on-site recovery of LFG, which is exported to the incineration plant through pipeline transportation. The incineration plant purchases the LFG to substitute coal as an alternative combustion promoter and evaporates the waste leachate. The SD model further hypothesizes that the purified LFG is provided by the landfill continuously, and there are no significant technical challenges to the leachate evaporation. Once the landfill is in the stage of closure, the integrated operation is expected to come to an end, and the incineration plant will return to using coal as the main combustion improver.

The sub-system of the LFG generation is based on the Marticorena Dynamics Model, which is a first-order dynamic equation, and has been widely applied to the prediction of LFG production [5, 6]. The model is given as follows (122, 126):

$$\mathrm{MP} = \mathrm{M}P_0 \cdot \exp(-\frac{t}{d}) \tag{5.8}$$

$$\mathrm{D(t)} = -\frac{dMP}{dt} = \frac{MP_0}{d} \cdot \exp(-\frac{t}{d}) \tag{5.9}$$

$$\mathrm{F} = \sum_i^t T_i \cdot D(t-i) = \sum_i^t T_i\left[\frac{MP_0}{d} \cdot \exp(-\frac{t-i}{d})\right] \tag{5.10}$$

where

MP	Landfill gas produced by municipal waste (MSW) degradation, m^3/tonne;
$MP0$	Landfill gas produced from fresh waste, m^3/tonne;
t	time, years;
d	time for duration of waste life, years;
$D(t)$	annual landfill gas production rate, m^3/y·tonne;
F	the landfill gas production rate of landfill, m^3/a;
Ti	the stock of waste in the ith layer, tonne

The sub-system of incineration plant operation mainly focuses on the related economic revenues. The main revenue comes from the power generation, municipal waste incineration, cost savings of combustion promoter, and government subsidies, whilst the main expenses are the cost of waste incineration, leachate treatment, LFG purchase, annual maintenance and construction investment of pipeline transportation.

The power generation from incineration is mainly determined by the calorific value of the waste [47, 92]. LFG can increase the waste calorific value, but the leachate sprayed into the incinerator may cause the decrease of calorific value. Thus, the purchased LFG α is ensured to a maintain a high waste calorific value. With regard to the leachate evaporation β, it should be set within an appropriate range. Otherwise, excessive leachate may result in reduction of calorific value and production of toxic pollutants.

5.2.2 An Illustrative Case

The Luodai municipal solid waste (MSW) incineration plant and the Chang'an solid waste sanitary landfill located in Chengdu, China are taken as a case study to obtain the data for the proposed SD model. The sanitary landfill is about 6 km away from

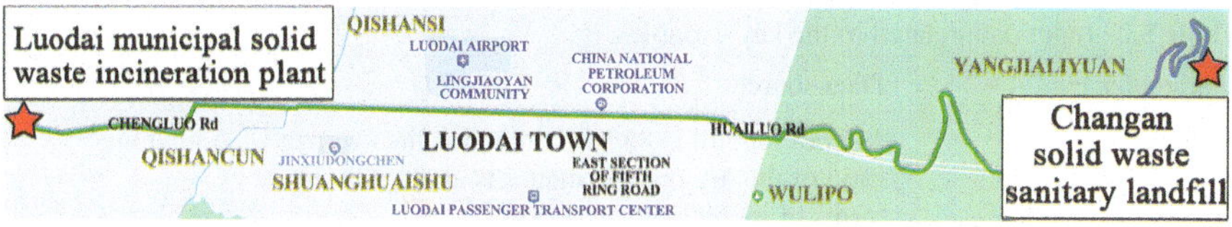

Fig. 5.5 Geographic location of the investigated incineration plant and landfill

the incineration plant, and their geographical locations are shown in Fig. 5.5. The designed treatment capacity of the Luodai incineration plant is 1,200 tons per day. There are about 300 m^3 of fresh leachate produced per day due to waste decomposition. The leachate is pre-treated on-site by physical method, and then transferred to a municipal sewage treatment plant for advanced treatment. Chang'an sanitary landfill has been receiving municipal waste since 1993, and has become the primary landfill for waste disposal in Chengdu. The landfill receives an average of 959,368 tons of waste annually, and the associated LFG has not yet been considered for recovery, but is emitted after torching, which poses a substantive risk to the public health and the regional environment [100].

The input data for the key variables are mainly derived from field surveys and relevant studies, which are listed in Table 5.2.

5.2.3 Model Validation

To ensure the model accuracy, it is necessary to verify the extent to which the model is consistent with the actual situation. In this study, behavior sensitivity is used for model validation, which identifies variables that are highly sensitive to the system. Moreover, the system behavior is examined to assess its reasonableness after calibrating the value corresponding to the identified sensitive variables [79, 87]. The ratio of the purchased LFG and the leachate evaporation ratio, which have a certain degree of subjectivity in the scenario analysis, are considered as two of the most critical variables affecting the economic performance of the incineration plant. The sensitivity is calculated by the slope between − 20% and + 20% of the variance of the two variables, to measure their influence on the annual revenue of the investigated incineration plant. The larger the slope is, the higher the sensitivity of the variables obtains.

By using the x-axis to indicate the purchased ratio of LFG, its change is examined to influence the annual revenue (indicated by the y-axis). The evaporation ratio decreased by 20% gives rise to an increase of 1.6% of the annual revenue, while the ratio increased by 20% results in a decrease of 1.5% in the annual revenue, as shown in Fig. 5.6.

Similarly, using the x-axis to indicate the evaporation ratio, its variance is also examined to influence the annual revenue. The purchased proportion of LFG decreased by 20% gives rise to an increase of 1.0% in the annual revenue, while a

Table 5.2 Input data related to the key variables

Parameter	Data source	Numerical Value
MP_0	MP_0 of landfill is 85 m^3/t, in which the average ratio of the dry organic matter is about 30% in terms of municipal waste [89]	85 m^3/tonne
d	The duration of landfill waste d is assigned as 4 years in terms of its stabilization, which is derived by analogy to Wang et al. [89]	4 years
T_i	By field investigation	1.02 million tonne/y
Cost of conventional leachate treatment	By field investigation	60 Yuan/tonne
Purchasing price of purified LFG	As there is no precise price of the purified LFG, it is thus converted by the industrial price of natural gas, set its benchmark as 3.14 Yuan/m^3 [88]. The conversion is depended upon the equivalent calorific value of purified LFG in contrast to the natural gas, in which the former is about 22,360 kJ/m^3, whilst the latter is 37,260 kJ/m^3	3.00 Yuan/m^3
Waste incinerated annually	The annual waste disposal by incineration is measured by the predicted population of Chengdu city multiplied by waste production per capita per day [42]	380,000–600,000 tonne/y
Governmental subsidy to the incineration plant	Due to a high maintenance cost of the incineration plant, local government is intended to give subsidy to support its operation [102]	180 Yuan/tonne
Annual utilization of combustion improver	The mass ratio of municipal waste and combustion improver is set as 0.8:0.2 [58]	100,000 tonne/y
The price of standard coal	The initial price is set by analogy to Cheng et al. [11]	310 Yuan/t
Feed-in Tariff of incineration	By field investigation	0.583 Yuan/kwh
Cost of waste incineration	By field investigation	70 Yuan/tonne
Storage ratio of fresh leachate	By investigation	20%
Cost of LFG pipeline construction	According to the field investigation, the construction fee for LFG pipeline is 500 Yuan/m, and the distance between landfill and incinerator is 6 km. Thus, the one-off investment is 3 million Yuan with 10 years lifespan of the pipeline. The annual fee equals to 300,000 Yuan	300,000 Yuan/y
Maintenance cost of incineration	By investigation	100 Yuan/tonne

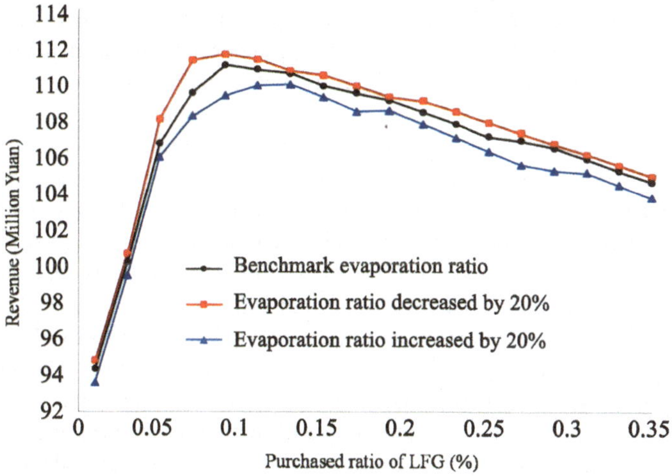

Fig. 5.6 Purchased ratio of LFG for sensitivity analysis

20% increase results in a 1.1% decrease in the annual revenue, as shown in Fig. 5.7. This result indicates the availability of the proposed model, which can rationally predict the outcomes when changes in system behavior occur.

Historical data comparison is also implemented for the behavioral validity. The common practice is to verify whether the simulation results of certain quantitative variables are consistent with the corresponding historical data [43]. This verification is performed by comparing the relative error between the historical data and the simulation results. However, the proposed system has not been implemented. It is

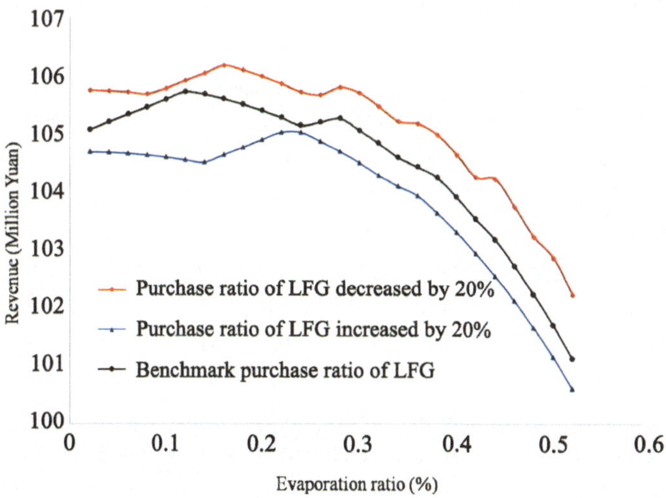

Fig. 5.7 Evaporation ratio for sensitivity analysis

difficult to compare the model outcomes directly with the actual economic performance. Especially, the filed survey reflects that the project manager is not likely to reveal the actual annual revenue of the investigated incineration plant, which has been commissioned since 2008. In this case, the study selects one of the key variables, i.e. landfill gas production, to validate the model by comparing its historical generation with the simulation result. As the investigated landfill was commissioned for waste disposal in 1993, the actual LFG production from 1995 to 2015 is in contrast with the simulation result, as shown in Table 5.3. The relative error of the tested variable is less than 18%, with an average error of 3.91%, which reflects that the simulation result is close to the actual generation, indicating that the proposed SD model is valid to reflect the real situation to some extent. Thus, further simulation can be conducted to investigate the possible variation in economic returns from changes in the purchased proportion of LFG and the proportion of leachate evaporation.

In order to investigate the possible impact of the purchased proportion of LFG α and the proportion of leachate evaporation β on the annual revenue of the investigated incineration plant, a benchmark revenue is determined by the independent operation of the municipal landfill and the incineration plant, as shown in Fig. 5.8. The simulation period of the established model was set to 30 years, which is consistent with the possible situation of LFG production. If the revenue of the integrated system is lower than the benchmark revenue, the attempt of interaction is considered inapplicable. In such a case, the incineration plant is still intended to use coal instead of LFG as the combustion improver. Additionally, if the calorific value is lower than the average waste calorific value of Chengdu City, i.e. 5631.6 kJ/kg offered by Yang [107], the leachate evaporation process is stopped.

The initial value of α is transformed by using the principle of heat conservation based on the purpose of substituting LFG for coal combustion improver. It has been calculated that 100,000 tons of standard coal can be substituted by 80,000 cubic metres of LFG [118, 119] when the integrated operation system is applied. Thus, the

Table 5.3 Behavioral validity based on historical data comparison

Year	1995	1996	1997	1998	1999	2000	2001
Simulation result (million m^3)	7.65	15.30	21.26	25.91	29.52	32.34	34.53
Historic data (million m^3)	7.91	13.58	17.97	22.62	25.50	30.35	33.71
Relative error (%)	3	13	18	14	16	7	2
Year	2002	2003	2004	2005	2006	2007	2008
Simulation result (million m^3)	36.24	37.57	38.60	39.41	40.04	40.53	40.91
Historic data (million m^3)	37.55	37.87	39.85	41.77	41.98	42.03	42.25
Relative error	3	1	3	6	5	4	3
Year	2009	2010	2011	2012	2013	2014	2015
Simulation result (million m^3)	41.20	41.44	41.62	35.79	31.26	27.73	24.98
Historic data (million m^3)	42.49	42.73	43.30	34.98	32.27	29.68	24.63
Relative error (%)	3	3	4	2	3	7	1

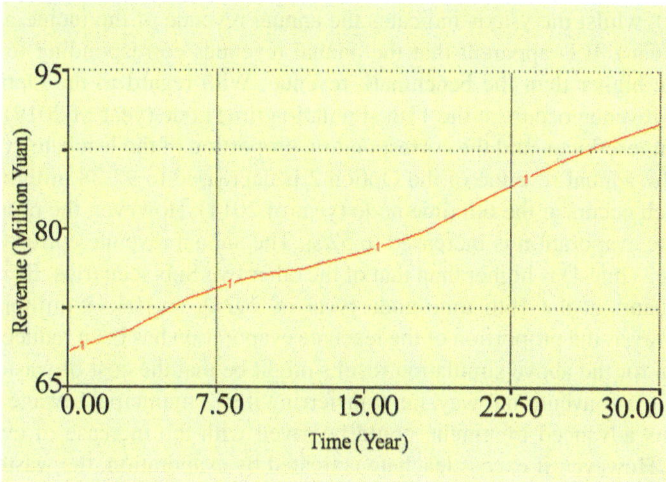

Fig. 5.8 Simulation result of the benchmark revenue

lower value of α is set as to 0.0042 based on the production of LFG in the year of 2014. For the evaporation ratio β, the initial value is set to zero due to the parallel treatment for the leachate. The interval of α and β is set to [0.0042, 1], and [0, 1] respectively. To find the optimal range of α and β, two scenarios are built by the division of the above intervals.

5.2.4 Results

In the scenario 1, the purchased proportion of LFG α is taken as a fixed value, while the proportion of leachate evaporation β is set by using the table function provided by the Stella Software Package. In order to understand the different influences on the annual revenue of the investigated incineration plant with the changes of β, three options are considered, i.e. ① the leachate evaporation keeps a constant growth rate, and considers β growing in a linear function, ② the growth of the evaporation rate decreases gradually, and considers β growing in a convex function, ③ the growth of the evaporation rate increases gradually, and considers β growing in a concave function. In each option, the interval of α is divided into 50 points by taking 0.02 as the step-length, fixed as 0.02, 0.04, 0.06, ultimately as 1.0. Each α is set as an input value for the SD model simulation, to seek for the maximum annual profit of the incineration plant, and the corresponding β. On this basis, the maximum annual profits of the three options are obtained, to seek for their corresponding purchased proportions of LFG α, by which the optimal LFG purchase proportion is determined.

By Taking $\alpha = 0.02$ as a typical example, Fig. 5.9 shows the SD simulation result of the annual revenue of the incineration plant. The x-axis represents the simulation time (year), whilst the y-axis indicates the annual revenue of the incineration plant (million Yuan). It is apparent that the annual revenues corresponding to the three options are higher than the benchmark revenue. With regard to the Option 1, the maximum revenue occurs at the 11th simulation time node (year of 2019), as 99.39 million Chinese Yuan, and the corresponding proportion of the leachate evaporation is 17%. The annual revenue of the Option 2 is decreased to 92.28 million Chinese Yuan, which occurs at the 6th time node (year of 2014). However, the proportion of the leachate evaporation is increased to 52%. The annual revenue corresponding to the Option 3 (line 4) is higher than that of the other two Sub-scenarios. Its maximum revenue occurs at the 16th time node (year of 2024), as 104.43 million Chinese Yuan. However, the proportion of the leachate evaporation has been reduced to 27%. The reason for the above simulation results might be that the cost of waste leachate treatment in a conventional way, i.e. transferring it to a municipal sewage treatment plant for its advanced treatment, could be saved with the increase of evaporation efficiency. However, if excess leachate is treated by evaporation, the waste calorific value in the incinerator may decrease dramatically, which gives rise to the loss of power generation and thus to reduce the revenue of the incineration plant.

In contrast to the Scenario 1, the proportion of leachate evaporation β is taken as a fixed value, whilst the purchased proportion of LFG α is set by different functions, i.e. linear, convex and the concave, which is also divided into three options to understand their influences on the annual revenue of the investigated incineration plant. In each option, the interval of β is divided into 50 points by taking 0.2 as the step-length,

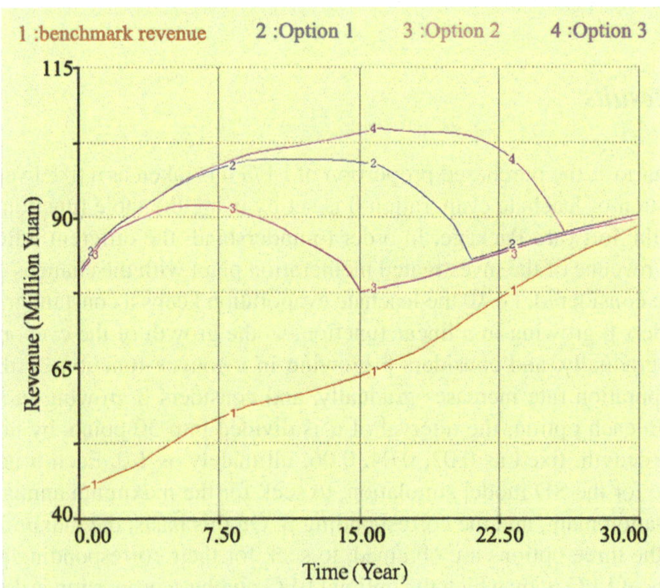

Fig. 5.9 Simulation result of the Scenario 1 by taking $\alpha = 0.02$ as an example

fixed as 0.02, 0.04, 0.06, ultimately as 1.0. Each β is set as an input value for the simulation, to seek for the maximum annual profit of the incineration plant, and the corresponding α. Taking $\beta = 0.2$ as a typical case example, three options are divided by analogy with the Scenario 1, i.e. ① the purchased proportion of LFG α keeps a constant growth rate, which takes α growth as a linear function, ② the growth of α decreases gradually, which takes α growth as a convex function, ③ the growth of α increases gradually, which takes α growth as a concave function.

Taking $\beta = 0.02$ as a typical example, Fig. 5.10 shows the SD simulation result of the annual revenue of the incineration plant. It is obvious that the annual revenues related to the Option 1 (line 2) and Option 3 (line 4), are superior to the benchmark revenue (line 1). In the Option 1, the maximum revenue occurs at the 6th simulation time node (year of 2014), as 100.07 million Chinese Yuan, and the corresponding proportion of the LFG purchase is 20%. The maximum revenue of the Option 3 is higher than that of the Option 1, as 105.06 million Chinese Yuan, occurs at the 9th simulation time node (year of 2017). However, its corresponding proportion of the LFG purchase is decreased by 10%. In contrast to the Option 1 and Option 3, it is clear that the annual revenue of the Option 2 (line 3) decreases dramatically, although there is a slight increase in the initial period. With its continuous decrease, it is completely below the benchmark revenue. The above phenomenon can be explained by the fact that the waste calorific value is improved by increasing the purchased proportion of LFG, which may further increase the capacity of leachate evaporation. However, excessive purchase of LFG may lead to financial burden on the economic success of the incineration plant. In such a case, the simulation result is indicated as failed to implement in actual situation, due to cost–benefit consideration.

With regard to the Scenario 1, i.e. the purchased proportion of LFG α is set as a fixed value, e.g. $\alpha = 0.02$, the annual revenue of the third option is apparently the largest among the three options. Furthermore, the proportion of the leachate evaporation β corresponding to the Option 3 is 10% higher than that of the Option 1. In case the annual revenue of the incineration plant of the Option 3 reaches the maximum as 110.92 million Yuan, the largest amount of the leachate evaporation is 41%. By taking both economic profit and leachate treatment into account, it is implied that the Option 3 is optimal for further management implementation. When α is input as 0.04, 0.06, 0.08…, etc., into the SD model, the similar results can be obtained, which indicates that the Option 3 is superior to the other two options.

aking $\beta = 0.02$ as an example in the Scenario 2, the annual revenue of the Option 2 is the lowest among the three options. Especially, it is even inferior to the benchmark profit at the 13th time node, which indicates that the Option 2 is not economically feasible for practical operation. Compared with the Option 3 and the Option 1, the former's revenue is higher than that of the latter, which can be considered as the optimal scenario for further management implementation. When β is input as 0.04, 0.06, 0.08…, etc., into the SD model, the similar simulation results can be obtained. In the case of $\beta = 0.12$, the maximum profit for the incineration plant is obtained as 105.71 million Yuan in the Option 3, with its optimal purchase ratio of LFG as 10%.

Based on the above analysis, the Option 3 in both the Scenario 1 and the Scenario 2 is identified as the optimal alternative for possible implementation. Figure 5.11

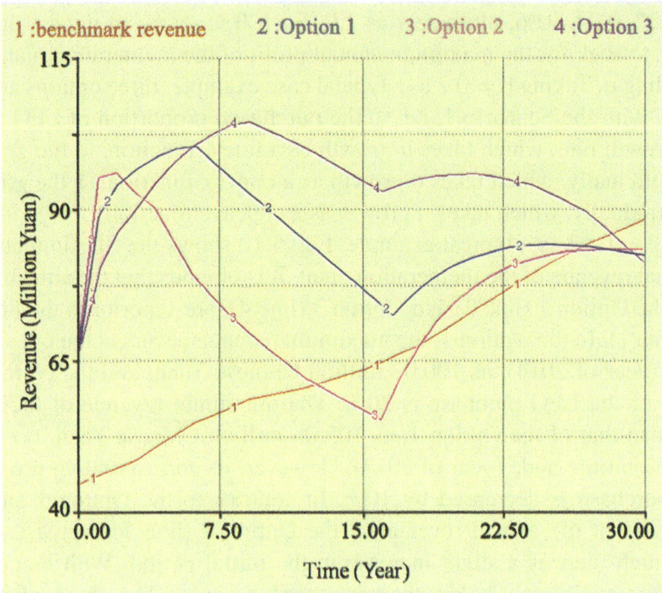

Fig. 5.10 Simulation result of Scenario 2 by taking $\beta = 0.02$ as an example

further demonstrates the variation of the annual revenue of the investigated incineration plant in relation to the two optimal options, which reflects that the optimal option of the Scenario 1 has a better economic performance than that of the Scenario 2.

As far as the treatment of leachate is concerned, the optimal option of the Scenario 1 is obviously superior to that of the Scenario 2, which is indicated by a larger

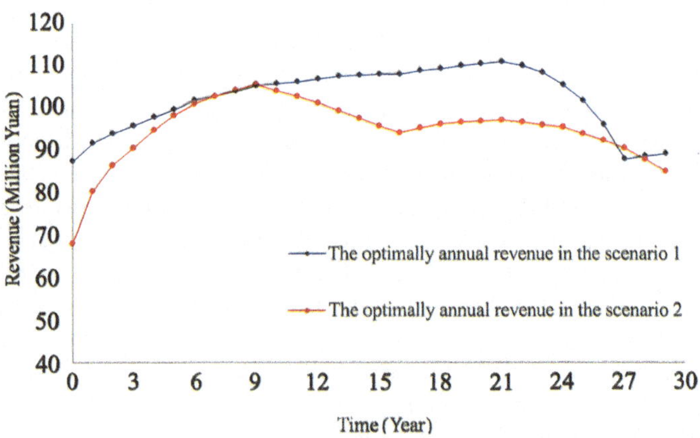

Fig. 5.11 Comparison of annual revenue between the two optimal options

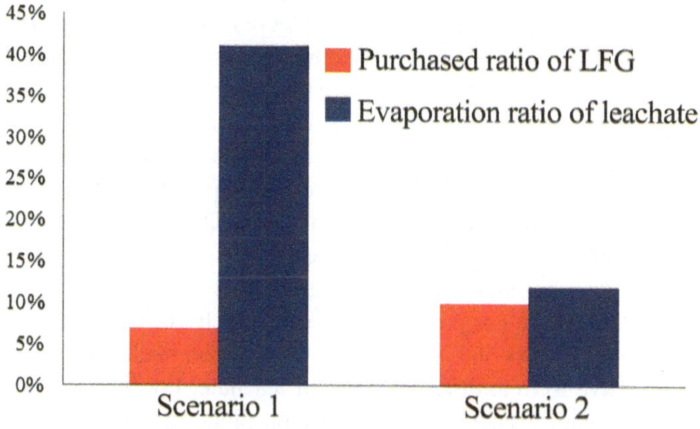

Fig. 5.12 Comparison of LFG purchase and leachate evaporation

proportion of the leachate evaporation, i.e. 4 times greater amount of the evaporation, as shown in Fig. 5.12.

Considering both the economic and environmental performances, Scenario 1 is deemed as the optimal operation scenario for further implementation, in which the incineration plant is capable of purchasing 7% of the purified LFG, and evaporating 41% of the on-site leachate. Ultimately, the annual revenue of the incineration plant is expected to be more than 110 million Yuan, 35% higher than the independent operation.

5.2.5 *Summary*

This Section offers a conceptual prototype by using LFG as an alternative energy source to treat the waste leachate. A system dynamics approach is applied to model the economic performance at a regional incineration plant. Two scenarios are established to demonstrate the application of the proposed SD model, set by changing the ratio of leachate evaporation and LFG utilization, to identify the optimal scenario with the best economic performance for future management implementation. The result shows that the Scenario 1 has a better economic performance than the Scenario 2, in which its revenue is 35% higher than the benchmark, as well as a maximum of 41% of the on-site leachate could be treated by the thermal evaporation. It is expected that the study can provide insight for the development of an alternative treatment mode of waste leachate and the reduction of the associated greenhouse gas emissions, thus improving sustainable waste management.

Although the results show that the integrated mode for leachate thermal treatment has a great potential for future implementation due to a substantial economic performance, there are still a number of uncertainties in evaluating the sustainable treatment system. With regard to the model, the uncertainty may be due to any of the quantitative expressions. For instance, the LFG in the model is hypothesized as a stabilized alternative energy source for the combustion substitute of the incineration plant. The actual LFG generation is a complex process that can be influenced by considerable factors, such as waste characteristics (density, biodegradability, moisture, etc.), micro-organisms content, aerobic or anaerobic conditions, the degree of initial compaction, etc. [9, 27, 106]. However, the empirical models for the prediction of the LFG production, such as the Marticorena Dynamics Model used in this study, the Mexico LFG model, the IPCC prediction model, the multi-phase prediction model, etc., may be not sufficient enough to discriminate whether the landfill is in the stage of intensive gas production or not, which may cause uncertainties of prediction [10, 90]. The second issue is that, evaporation itself, as a single process, may not be effective to treat different types of leachate. For instance, the relationship between leachate dosage in the incineration plant and evaporation efficiency may bring uncertainties in simulation results. There needs a careful design regarding leachate evaporation, particularly for the associated volatile component control and reduction of high inorganic salt concretion on the incinerator furnace wall [118]. It is thus recommended to combine the evaporation with other alternative treatment options. Specifically, leachate is generally divided into the young and mature leachate, according to the age of the landfill [59]. The former may have higher BOD and COD contents than the latter, so the biological process coupled with evaporation is expected to have a good performance in the leachate treatment [51]. With regard to the mature leachate, due to its low biodegradability, refractory organics consisting of dissolved organic matter (DOM), evaporation integrated with reverse osmosis (RO), ultrafiltration, etc., are recommended as a solution to reduce its organic and ammonia level [92]. From the perspective of data uncertainty, it may come from its acquisition and reliability, e.g. there is a lack of statistical data related to the economic performance of the investigated incineration plant because it is newly built, so the model may not be robustly validated.

5.3 Game Theoretical Analysis to the Conflicts of Waste Disposal

Rapid urbanization has witnessed increasing generation of municipal solid waste (MSW) in developing countries [61]. China, as a developing country with large population, produces more than 400 million tons of MSW annually, resulting in a remarkable issue as "besieged by waste" [81]. To mitigate such consequence, incineration is taken as an important alternative for waste treatment, which is not only beneficial for waste reduction, but also for energy recovery [103]. By the end of 2019, 418 incineration plants have been constructed in China [85, 103]. However, incineration faces considerable protests from local communities, due to its negative externalities, such as odor emissions, dioxins in its flue gas etc., resulting in a series of not-in-my-backyard (NIMBY) conflicts, including public petitions, street marches, etc. [68]. The NIMBY conflict indicates a contradictory phenomenon that the public recoginizes that a built municipal facility can bring social benefits, but still opposes its construction if it is close to their homes [78].

A possible reason for the NIMBY conflicts is believed by the public risk perception, due to their limited knowledge and biased consciousness, who may increase the related impacts, such as degradation of quality of life and environment, form a motive to oppose on the construction and operation of NIMBY facilities [16, 116]. With regard to the incineration projects, more than ten anti-incinerator campaigns have occurred in the last decade, calling for the relocation of incineration facilities [77].

In addition to the public, the local government is considered another sensitive stakeholder involved in the conflicts, since its intention on positive administrative practice can bring environmental and social benefits to communities that further reduces public risk perception [66]. The Chinese government has formulated a series of regulations on waste incineration projects, such as the revised "Regulations of the People's Republic of China on the Disclosure of Government Information", the revised "Standard for Pollution Control on the Municipal Solid Waste Incineration", "Measures on Public Participation in Environmental Impact Assessment", etc [37]. These measures aim at improving the supervision on the operation of incineration plants, e.g. the pollutant emissions standards have been significantly improved, some of which are even stricter than the corresponding EU standard, in order to ensure public appeal on maintaining health safety and environmental quality [67]. Although the operational enterprises of incineration plant are required to comply with these regulations, they are not willing to increase capital investment in technological upgrading to further improve the environmental quality of their workplaces. This is common in China's industries of waste treatment and disposal, since they are considered public welfare sectors whose revenues are mainly come from government subsidies [49].

Given the above interactions, it is implied that there are a number of interests and conflicts among the public, local government, and the operations enterprise of the incineration plant, to form a three-player game. For the enterprise, which is motivated by profit, it is unwilling to take more corporate social responsibility, while it has to act in accordance with the basic policy constraints [120]. The public, especially the residents near the incineration facility, may perceive the possible change in environmental quality to increase their subjective risk [39, 76]. In any case, if they have not benefitted from such an incineration project, including employment opportunities, transportation convenience, environmental improvement, etc., it may trigger their protests for compensation from the enterprise [46]. The government is delicated to improving environmental management to sanction the environmentally unfriendly behaviour on the one hand, and to fully understand the publics' appeal on the other hand, ultimately to create a sustainable society [15, 46]. Game theory will allow a strategic model to be examined through the above actions, in order to provide an equilibrium strategy to mitigate possible conflicts between the three stakeholders.

As a classical game is generally limited by its assumption of complete rationality of the involved players [4], this study constructs an evolutionary game to investigate the transformation of the players' strategic actions, and thus identify the appropriate strategic action for alleviating the conflicts regarding the operation of waste incineration. System Dynamics (SD) is used to test such actions and predict their future development. The joint application of evolutionary game theory and system dynamics can maximize their advantages, facilitating the study of the dynamic change of equilibrium strategy [124]. The potential findings are expected to better understand the interactions of multiple stakeholders involved in the waste incineration conflicts, and provide insightful perspectives on policy implications in mitigation of such conflicts.

5.3.1 Game Model Formulation

The static Nash equilibrium defined by the von Neumann and Morgenstern cannot explain the dynamic interactions of stakeholders in conflicts, which facilitates the development of evolutionary game [71]. By adopting the natural selection mechanism instead of the strict rational assumption, evolutionary game can reasonably reflect the strategy evolution process of stakeholders with different interests , making it an important and powerful tool to study the dynamic characteristics of multi-group game behavior [112].

The strategic actions of each player are not only influenced by its own decisions, but also by those of other players. For the enterprise, it has two strategic options: one is to upgrade and optimize its treatment facilities to improve environmental quality, including odour reduction, purification of flue gas etc. (UT), and the other is to make no change to the existing facilities (NUT). The local government has two strategic options, indicating that one is to conduct regular inspection on the operation of incineration plant (RI), while another one is to conduct irregular inspection (IRI). With regard to the residents nearby, their strategic options are discriminated by whether they appeal or not to compensation for their quality of life and environment (AC, NAC).

According to the above strategic actions, the game proposes the following assumptions:

(1) All the involved players are hypothesized as rational economic agents with bounded rationality, and their decision making is based upon maximizing their own interests.
(2) The probability that the enterprise does not upgrade treatment facilities is given as p_1, whilst the probability of upgrading is 1 - p_1. The probability that the local government imposes regular inspection on the incineration plant is p_2, and the probability of irregular inspection is $1 - p_2$. The probability that the residents will demand compensation is p_3, and the probability that they will not demand compensation is $1 - p_3$.

The specific designed parameters to compose of the game, and their interpretations are given in Table 5.4.

Table 5.4 Model parameters

Parameter	Interpretation
B_a	Enterprise's income from power generation
B_b	Enterprise' income from waste disposal
A_e	Operation and maintenance cost of incineration plant
A_{eu}	Cost of treatment facilities upgrade
A_{g1}	Social cost
A_{g2}	Inspection cost
A_p	Cost of residents appealing to compensation
T	Tax revenue
D_p	Compensation for residents
D_g	Administrative penalty by the enterprise
E	Health loss associated with typical diseases of residents

The payoffs of the three players with their different strategic options are shown in Table 5.5.

According to the payoff matrix among the three players, P_1, P_2 and P_3 are used to denote the expected payoff functions of the enterprise, government and residents, respectively.

$$P_1 = p_1(B_a + B_b - A_e - T - p_2 D_g - p_3 D_p) + (1 - p_1)(B_a + B_b - A_e - A_{eu} - T) \tag{5.11}$$

$$P_2 = p_2(p_1 D_g - p_3 A_{g1} + T - B_b - A_{g2}) + (1 - p_2)(T - B_b - p_3 A_{g1}) \tag{5.12}$$

Table 5.5 Payoffs matrix for the three players

Enterprise	Government	Residents	
		AC (p_3)	NAC ($1 - p_3$)
NUT (p_1)	RI (p_2)	$(B_a + B_b - A_e - T - D_g - D_p$ $T + D_g - A_{g1} - A_{g2} - B_b$ $D_p - E - A_p)$	$(B_a + B_b - A_e - T - D_g$ $T + D_g - A_{g2} - B_b$ $- E)$
	IRI ($1 - p_2$)	$(B_a + B_b - A_e - T - D_p$ $T - A_{g1} - B_b,$ $D_p - E - A_p)$	$(B_a + B_b - A_e - T$ $T - B_b$ $-E)$
UT ($1 - p_1$)	RI (p_2)	$(B_a + B_b - A_e - A_{eu} - T$ $T - B_b - A_{g2} - A_{g1}$ $-A_p)$	$(B_a + B_b - A_e - A_{eu} - T$ $T - B_b - A_{g2}$ $0)$
	IRI ($1 - p_2$)	$(B_a + B_b - A_e - A_{eu} - T$ $T - B_b - A_{g1}$ $-A_p)$	$(B_a + B_b - A_e - A_{eu} - T$ $T - B_b$ $0)$

$$P_3 = p_3(p_1 D_p - A_p - p_1 E) + (1 - p_3)(-p_1 E) \tag{5.13}$$

Thus, the replication dynamic equations $F(p_i)$ corresponding to the enterprise $F(p_1)$, government $F(p_2)$ and residents $F(p_3)$ are expressed as follows

$$F(p_1) = \frac{dp_1}{dt} = p_1(1 - p_1)(A_{eu} - p_2 D_g - p_3 D_p) \tag{5.14}$$

$$F(p_2) = \frac{dp_2}{dt} = p_2(1 - p_2)(p_1 D_g - A_{g2}) \tag{5.154}$$

$$F(p_3) = \frac{dp_3}{dt} = p_3(1 - p_3)(p_1 D_p - A_p). \tag{5.16}$$

5.3.2 Game Theoretical Analysis

The replication dynamics equation reflects the speed and direction of evolution with respect to the involved players [111]. When $F(p_i) = 0$, the game reaches a relatively stable equilibrium state. Let $F(x) = (F(p_1), F(p_2), F(p_3))T = 0$, the Nash Equilibrium of the evolutionary game can be obtained as follows:

$x_1 = \begin{cases} 0 \\ 0 \\ 0 \end{cases}$	$x_2 = \begin{cases} 0 \\ 0 \\ 1 \end{cases}$	$x_3 = \begin{cases} 0 \\ 1 \\ 0 \end{cases}$
$x_4 = \begin{cases} 0 \\ 1 \\ 1 \end{cases}$	$x_5 = \begin{cases} 1 \\ 0 \\ 0 \end{cases}$	$x_6 = \begin{cases} 1 \\ 0 \\ 1 \end{cases}$
$x_7 = \begin{cases} 1 \\ 1 \\ 0 \end{cases}$	$x_8 = \begin{cases} 1 \\ 1 \\ 1 \end{cases}$	$x_9 = \begin{cases} p_1^* \\ p_2^* \\ p_3^* \end{cases}$

Specifically, x_9 is the solution of

$$\begin{cases} A_{eu} - p_2 D_g - p_3 D_p = 0 \\ p_1 D_g - A_{g2} = 0 \\ p_1 D_p - A_p = 0 \end{cases} \tag{5.17}$$

The stability of the above equilibrium points can be judged by the Jacobian matrix via local stability analysis [23]. When the equilibrium points satisfy the matrix determinant $det(J) > 0$ and the matrix trace $tr(J) < 0$, they are defined as the asymptotically stable fixed points, which is called the evolutionary stability strategy [23, 124].

$$F_{11} = \frac{\partial F(p_1)}{\partial p_1} = (1 - 2p_1)(A_{eu} - p_2 D_g - p_3 D_p) \tag{5.18}$$

$$F_{12} = \frac{\partial F(p_1)}{\partial p_2} = -D_g p_1 (1 - p_1) \tag{5.19}$$

$$F_{13} = \frac{\partial F(p_1)}{\partial p_3} = -D_p p_1 (1 - p_1) \tag{5.20}$$

$$F_{21} = \frac{\partial F(p_2)}{\partial p_1} = D_g p_2 (1 - p_2) \tag{5.21}$$

$$F_{22} = \frac{\partial F(p_2)}{\partial p_2} = (1 - 2p_2)(p_1 D_g - A_{g2}) \tag{5.22}$$

$$F_{23} = \frac{\partial F(p_2)}{\partial p_3} = 0 \tag{5.23}$$

$$F_{31} = \frac{\partial F(p_3)}{\partial p_1} = p_3 (1 - p_3) D_p \tag{5.24}$$

$$F_{32} = \frac{\partial F(p_3)}{\partial p_2} = 0 \tag{5.25}$$

$$F_{33} = \frac{\partial F(p_3)}{\partial p_3} = (1 - 2p_3)(p_1 D_p - A_p) \tag{5.26}$$

$$J = \begin{bmatrix} F_{11} & F_{12} & F_{13} \\ F_{21} & F_{22} & F_{23} \\ F_{31} & F_{32} & F_{33} \end{bmatrix} = \begin{bmatrix} (1 - 2p_1)(A_{eu} - p_2 D_g - p_3 D_p) & -D_{g1}(1 - p_1) & -D_p p_1 (1 - p_1) \\ D_g p_2 (1 - p_2) & (1 - 2p_2)(p_1 D_g - A_{g2}) & 0 \\ p_3 (1 - p_3) D_p & 0 & (1 - 2p_3)(p_1 D_p - A_p) \end{bmatrix} \tag{5.27}$$

According to the Lyapunov's method, the equilibrium point is an ESS, only if all the eigenvalues in the Jacobian matrix have negative real parts [23, 97]. If all the eigenvalues in the Jacobian matrix are positive, the equilibrium point is considered to be an unstable point [23, 97]. If there are one or two positive numbers in the eigenvalues of the Jacobian matrix, the equilibrium point is considered to be a saddle point [23, 94, 97]. In this study, we only test the equilibrium points x_1 to x_8, as the solution to a three-player evolutionary game is typically a strict Nash equilibrium [44]. The eigenvalues for the equilibrium points x_1 to x_8 are given in Table 5.6.

Table 5.6 The main eigenvalues for the equilibrium points

(p_1, p_2, p_3)	F_{11}	F_{22}	F_{33}
(0,0, 0)	A_{eu}	$-A_{g2}$	$-A_p$
(0,0, 1)	$A_{eu} - D_p$	$-A_{g2}$	A_p
(0,1, 0)	$A_{eu} - D_g$	A_{g2}	$-A_p$
(0,1, 1)	$A_{eu} - D_g - D_p$	A_{g2}	A_p
(1,0, 0)	$-A_{eu}$	$D_g - A_{g2}$	$D_p - A_p$
(1,0, 1)	$-A_{eu} + D_p$	$D_g - A_{g2}$	$-D_p + A_p$
(1,1, 0)	$-A_{eu} + D_g$	$-D_g + A_{g2}$	$D_p - A_p$
(1,1, 1)	$-A_{eu} + D_g + D_p$	$-D_g + A_{g2}$	$-D_p + A_p$

Note $F_{12} = F_{13} = F_{21} = F_{23} = F_{31} = F_{32} = 0$

Since $A_{eu} > 0$, $A_p > 0$ and $A_{g2} > 0$, the equilibrium points (0,0, 0), (0,0, 1), (0,1, 0) and (0,1, 1) can be unstable points or saddle points. The equilibrium points (1,0, 0), (1,0, 1), (1,1, 0) and (1,1, 1) can be the potential evolutionary stability strategies. Among these potential ESS, the strategy that the residents appeal not to compensation (NAC) is the ideal stability strategy.

The equilibrium point (1,0, 0) is the ESS, when $D_g - A_{g2} < 0$ and $D_p - A_p < 0$. The equilibrium point (1,0, 0) is a saddle point, when $-A_{eu}$, $D_g - A_{g2}$ and $D_p - A_p$ have one or two positive numbers. Under the conditions that enterprise keeps no change to the existing facilities (NUT) as well as residents appeal no compensation, the local government can adopt irregular inspection on the operation of incineration plant. The equilibrium point (1,1, 0) is discussed similarly. When $D_g - A_{eu} < 0$, $A_{g2}-D_g < 0$ and $D_p - A_p < 0$, the equilibrium point (1,1, 0) is the ESS. When $D_g - A_{eu} > 0$, $A_{g2}-D_g > 0$ and $D_p - A_p > 0$, the equilibrium point (1,1, 0) is an unstable point. When $D_g - A_{eu}$,$A_{g2}-D_g$ and $D_p - A_p$ have one or two positive numbers, the equilibrium point (1,1, 0) is a saddle point. The results indicate that the enterprise adopts strategic action as NUT when the cost of upgrading the treatment facilities is higher than the administrative penalty. The government imposes "regular inspection" (RI) action when its inspection cost is less than the administrative penalties imposed by the enterprise. The residents choose not calling for compensation (NAC), if the compensation they receive is less than their costs.

5.3.3 System Dynamics Simulation

In order to reflect the dynamic evolution of the strategic actions, the study adopts SD to investigate the game evolution process, thus searching for the possible stability strategies. The study uses Vensim software package to build the SD model. The SD model in this study is composed of three subsystems, namely the enterprise subsystem, the government subsystem, and the residents subsystem, respectively. The model consists of 3 level variables, 3 flow variables, 10 external variables, and

21 intermediate variables, as shown in Table 5.7. The level variables represent the probability of strategic actions that the involved stakeholders can choose, including the actions related to the enterprise, government, and residents, respectively. The flow variables represent the rate of change corresponding to the probability of the strategic actions. The external variables have been claimed in the hypotheses of the proposed game model. The intermediate variables are the possible influencing factors related to the stakeholders' payoffs, which come from similar studies and field investigation. For example, Song et al. [74] pointed out that the polluting enterprise should be responsible for providing the necessary environmental compensation to the affected residents nearby to reduce the possible health damage. At the same time, the residents can perceive the potential environment impacts resulting from the enterprise's operation and negotiate for compensation for their quality of life [94]. Besides, government supervision and incentives may be the drivers to compel the enterprise to improve environmental quality, thus responding to the public's concern about environmental health [26, 70].

For the enterprise, the expected payoffs are mainly reflected in whether to upgrade facilities or not, which is determined by the profit from power generation, the cost of waste incineration, the cost of upgrading facilities and administrative penalties, as shown in Fig. 5.13a. The improvement of environmental quality by upgrading treatment facilities may bring marginal costs [45]. In such case, the enterprise may be unwilling to take such action unless the possible benefits cover the costs, i.e. the expected payoff is above 0. Based on the above analysis, there are two main causal loops in the enterprise subsystem.

Reinforcing loop: Probability that the enterprise upgrades facilities → Health loss that residents exposed to environment → Administrative penalties → Enterprise's expected payoffs with upgrading facilities → Probability that the enterprise upgrades facilities;

Balancing loop: Probability that the enterprise upgrades facilities → Health loss that residents exposed to environment → Expected payoffs of governmental regular inspection → Probability that government implements regular inspection → Probability that the enterprise upgrades facilities.

The payoffs of the governmentsubsystem are determined by the costs of regular inspections, administrative penalties from the enterprise and emergency costs for mass unexpected incidents, as shown in Fig. 5.13b. The primary responsibility of the local government is to encourage the enterprise to comply with environmental regulations either through economic policy instruments such as tax preference, subsidies, etc., or through administrative sanctions [26, 70, 120]. At the same time, the local government assumes the responsibility of understanding the public's appeal to improve environmental quality to avoid possible group conflicts [12, 101]. There are two main causal loops in the government subsystem, which are defined as follows.

Reinforcing loop: Expected payoffs of governmental regular inspection → Probability that government implements regular inspection → Probability that the enterprise upgrades facilities → Health loss that residents exposed to environment → Administrative penalties → Expected payoffs of governmental regular inspection;

Table 5.7 Variable type of the SD model

Variable	Type	Variable	Type	Variable	Type
Administrative penalties	Auxiliary	Expected payoffs of residents who do not appeal to compensation	Auxiliary	Probability change on residents	Rate
Compensation factor	Constant	Expected payoffs of residents who appeal to compensation	Auxiliary	Probability that government implements regular inspection	Level
Concentration of pollutants	Constant	Expected payoffs that the government does irregular inspection	Auxiliary	Probability that residents appeal to compensation	Level
Cost for enterprise to upgrade facilities	Auxiliary	Expected payoffs of government	Auxiliary	Probability that the enterprise upgrades facilities	Level
Cost of complaints	Auxiliary	Government subsidy	Constant	Profit of power generation	Auxiliary
Difference in profit while upgrading facilities	Auxiliary	Health loss that residents exposed to environment	Auxiliary	Residents' compensation	Auxiliary
Difference in payoffs under governmental regular inspection	Auxiliary	Local salary	Constant	Standard value of pollutants	Constant
Economic penalty	Constant	Maintenance cost	Auxiliary	The cost of governmental regular inspection	Constant
Enterprise's expected payoffs	Auxiliary	Expected payoffs of residents	Auxiliary	The expected compensation of residents	Auxiliary
Enterprise's expected payoffs with upgrading facilities	Auxiliary	Power generation	Constant	Cost of waste incineration	Auxiliary
Enterprise's expected payoffs without upgrading facilities	Auxiliary	Power generation capacity	Constant	Waste treatment capacity	Constant
Illegal emissions	Auxiliary	Probability change on enterprise	Rate		
Expected payoffs of governmental regular inspection	Auxiliary	Probability change on government	Rate		

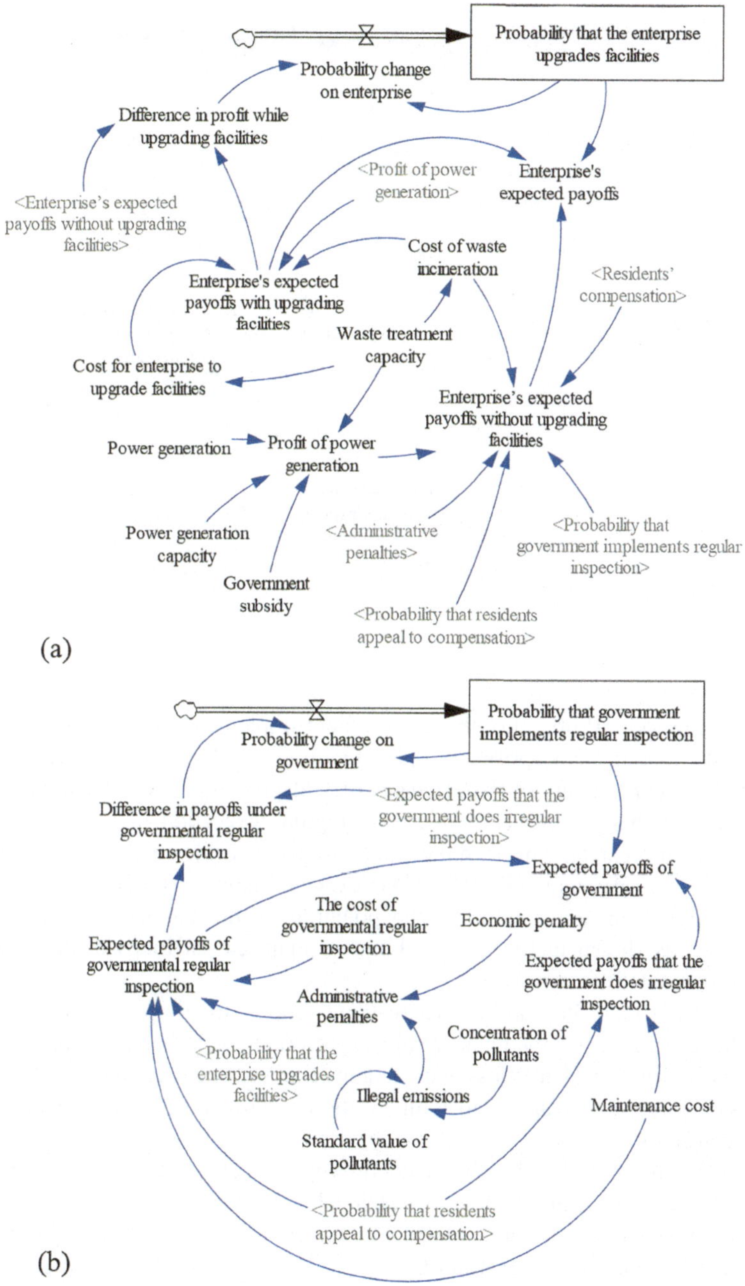

Fig. 5.13 The three players' evolutionary game-based SD model. **a** The enterprise. **b** The government. **c** The residents

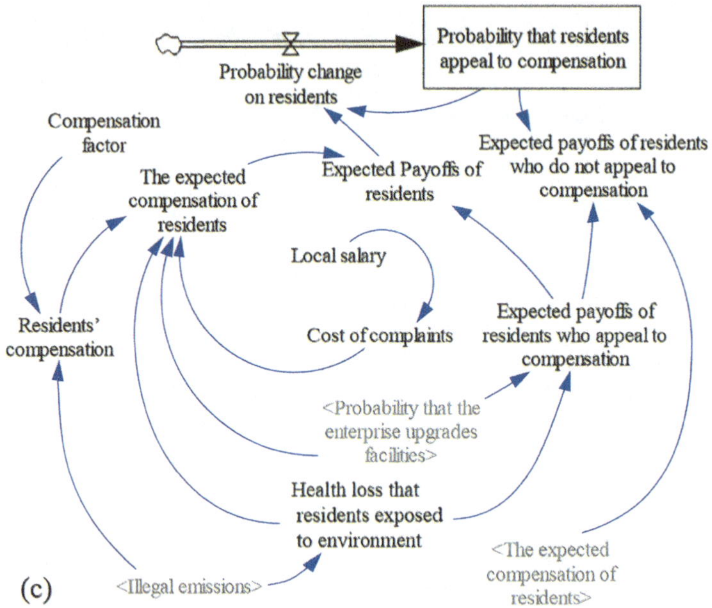

Fig. 5.13 (continued)

Balancing loop: Probability that government implements regular inspection → Probability that the enterprise upgrades facilities → Health loss that residents exposed to environment → Expected payoffs of governmental regular inspection → Probability that government implements regular inspection.

The residents can bargain with the enterprise for possible compensation for their perceived health damages [32, 94]. Their expected payoffs are determined by the compensation from the enterprise, the potential health damage and the cost of complaints, as shown in Fig. 5.13c. The two main causal loops in the residents subsystem are defined as follows:

Reinforcing loop: Probability that residents appeal to compensation → Probability that government implements regular inspection → Probability that enterprise upgrades facilities → Health loss that residents exposed to environment → Expected payoffs of residents who appeal to compensation → Probability that residents appeal to compensation

Balancing loop: Probability that residents appeal to compensation → Probability that the enterprise upgrades facilities → Health loss that residents exposed to environment → Expected payoffs of residents who appeal to compensation → Probability that residents appeal to compensation.

5.3.4 An Illustrative Case

In this section, a case study is provided to verify the application of the proposed three-player game. A municipal incineration plant located in Nanchong City, Southwestern China, which has received great public attention, is selected as a case study. The plant was completed in 2018 with a maximum processing capacity of 800 tons per day. According to the field survey, only 40 households live within a 1-km radius of the plant, and the rest have been relocated. It is indicated that 93% of the respondents feel depressed by the operation of the incineration plant, e.g. worried about their health and crops due to the odour and smoke. The input parameters for the SD model are mainly obtained from field surveys and similar studies, as shown in Table 5.8.

5.3.5 Results

According to the Sect. 3.5.2 as the analysis of evolutionary stability strategy, when $D_g - A_{eu} < 0$, $A_{g2} - D_g < 0$ and $D_p - A_p < 0$, the game finally achieves an evolutionary stability, and the equilibrium point (1,1, 0) corresponds to the pair of strategic actions as [NUT, RI, NAC]. In this case, the initial probability that the enterprise prefers not to upgrade the treatment facilities is set to $p_1 = 0.7$, the initial strategy that the government imposes regular inspection as $p_2 = 0.5$, and the initial strategy that the residents claim environmental compensation as $p_3 = 0.5$, respectively. Figure 5.14 shows that the three players can follow by stable strategic actions after a short simulation period (about 5–15 years), indicating that the proposed game has stable equilibrium points.

It is clear that the probability of NUT $1 - p_1$, the probability of RI p_2, and the probability of AC p_3 all converge to a stable strategic equilibrium point sets as (0.66, 1, 0), respectively. Such a phenomenon implies dominant strategies for both the local government and the residents, i.e. the local government will impose regular inspection on the enterprise's operational behaviour, whilst the residents choose not to seek compensation.

For the enterprise, it has the potential to upgrade its treatment facilities to solve the issues related to odour emissions, as its strategic probability is slightly above 0.5 (shown in Fig. 5.14a). Among the three players, the probability of government action converges the slowest (shown in Fig. 5.14b), while the residents' strategic action converges the fastest (shown in Fig. 5.14c).

The above simulation results indicate that the enterprise has a probability of 0.66 to upgrade its treatment facilities, which leaves room for further improvement. In such a case, this study constructs an incentive scenario and a sanction scenario to compare their performance on the strategic actions of the three players.

As the cost and benefit factor is the key that drives the enterprise to implement technological innovation or not, the incentive scenario mainly on providing government subsidies to cover the cost of upgrading the facilities of deodorization. Figure 5.15

Table 5.8 Interpretation of the key influencing factors

Parameter	Value	Interpretation
Enterprise's income from power generation (B_a)	–	The income from power generation can be measured by generating capacity multiplying by electricity price, in which the electricity price after government subsidy is 0.65 CNY/Kwh [60]
Operation and maintenance cost of incineration plant (A_e)	–	The operation and maintenance cost is derived from similar study, as in a range of 75 to 120 CNY/t [19]
Cost of upgrading treatment facilities (A_{eu})	16 CNY/t	The unit cost of waste disposal is increased by 16 CNY/t since the treatment facilities upgrading [52]
Social cost (A_{g1})	–	It is hypothesized as the expense for government in mediation of the conflict, which can be expressed as: $T \times A$ where T is consistent with the time that the residents spent on appealing to their compensation A is the basic allowance issued by the Ministry of Finance, i.e. each one will obtain 180 CNY per day [54]
Cost of residents appealing to compensation (A_p)	–	Since there is no specific measurements on the cost of complaints, this study estimates it as: $T \times S$ where T is the time that the residents spent for appealing to their compensation S is the local average salary, its value is 192 CNY/person/day [75]
Compensation for the residents (D_p)	–	The pollution compensation for residents nearby can be assessed as [109]: $K \sum_{i=1}^{n} \max\left(\frac{X_i(x,y,o)}{\overline{x_t}} - 1,0\right)$ where K is the compensation coefficient, assumed as 2000 [109] $X_i(x, y, 0)$ is the concentration of the ith pollutant at the geographic location $(x, y, 0)$ $\overline{x_t}$ represents the emissions standard of the ith pollutant, selected as 80 mg/Nm^3 [53]
Administrative penalty (D_g)	–	The penalty can be measured as followed [63]: $\frac{X_i(x,y,o)}{\overline{x_t}} \times 100{,}000$
Health loss associated with typical diseases (E)	424.81 CNY	The measurement is given as follows [107] $\sum_{i=1}^{3} (PAR_i \cdot E_i \cdot P_i)$ where PAR_i is the different types of diseases caused by pollution, including chronic bronchitis, cor-pulmonale, lung cancer etc E_i is the average cost for the diagnosis and medical treatment P_i is the probability of a person suffered by the diseases
Treatment capacity	219,000 t	According to the field investigation
Generating capacity per ton of waste	345 Kwh/t	According to the field investigation

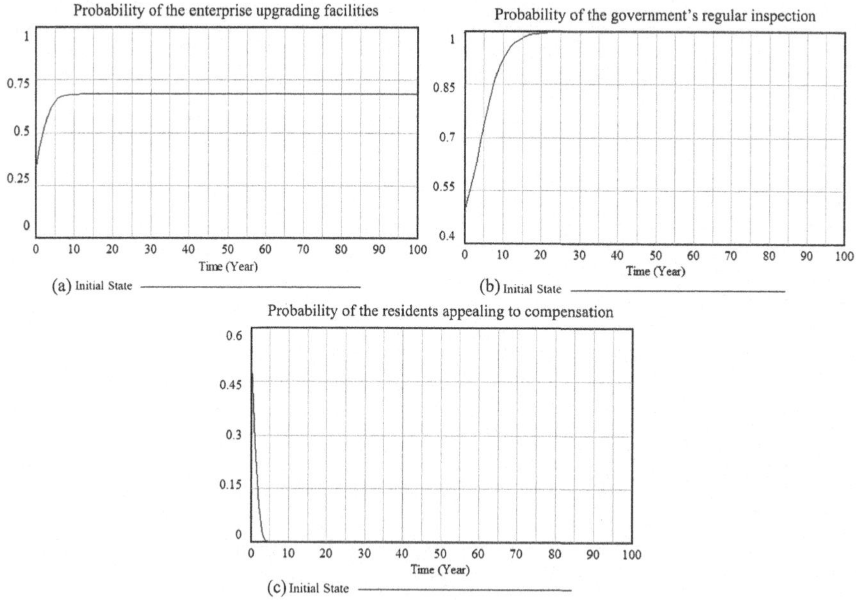

Fig. 5.14 Simulation results of the proposed game. **a** Probability of the enterprise upgrading facilities. **b** Probability of the government's regular inspection. **c** Probability of the residents appealing to compensation

shows the variation of strategic actions corresponding to the three players. It can be seen that the strategy equilibrium points have not varied for the three players, but their rates of convergence are apparently different. The probability of the enterprise's action converges the fastest (shown in Fig. 5.15a), indicating that government subsidies have an incentive effect on the enterprises' behaviour. However, Fig. 5.15b shows that the probability convergence corresponding to the government's strategic action is slowerbecause the subsidy to some extent reduces the government's expected payoffs. The probability convergence of the residents' action shows the least change (shown in Fig. 5.15c), indicating that the government incentives have less impact on their decision making.

Moreover, the equilibrium strategies of the three players have not been significantly improved. It is thus implied that the government incentives may be not efficient in motivating the enterprise for technical transformation [21]. Fan and Dong [21] have found that the performance of incentives on technological innovation of enterprises is uncertain. In fact, whether an enterprise can undertake technological innovation depends not only on the capital investment, but also on its research and development capability and the prediction of market competition.

The second scenario is that the local government imposes an economic sanction on the enterprise for illegal behavior, such asodour emissions without taking any countermeasures, to force the enterprise to facilitate the upgrade of treatment facilities. Figure 5.16 a shows that the probability of choosing to upgrade has significantly

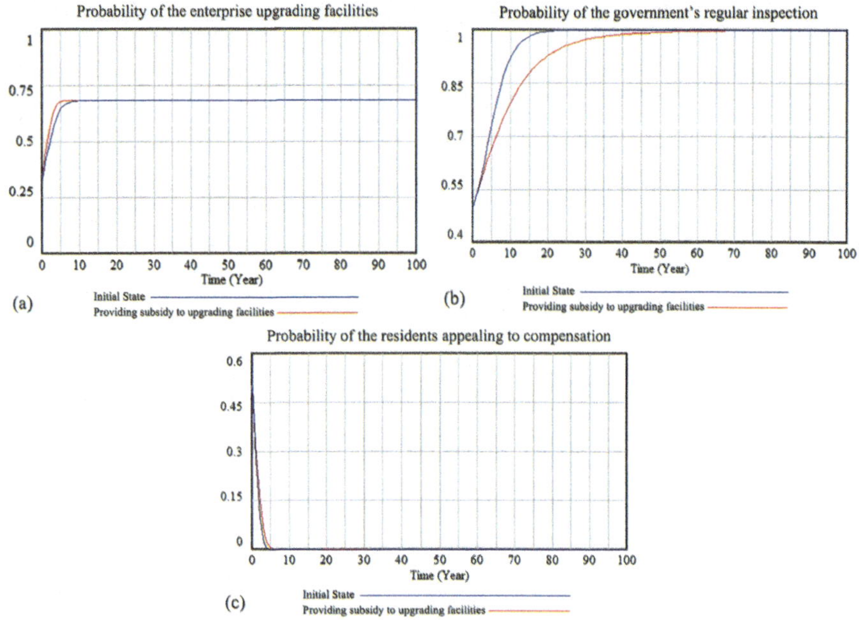

Fig. 5.15 Simulation results by government providing subsidy incentives. **a** Probability of the enterprise upgrading facilities. **b** Probability of the government's regular inspection. **c** Probability of the residents appealing to compensation

increased from 0.66 to 0.82. Although the final probability for government action is 1, its rate of convergence has been improved, indicating that the local government will strengthen the intensity of supervision, as shown in Fig. 5.16b. Figure 5.16c shows that the probability equilibrium of residents' action remains almost the same as 0. It thus implies that reinforcement of inspection by increasing administrative penalties is an effective way to impel the enterprise to improve its environmental quality, by which the residents can satisfy its operation without resorting to compensation. A similar indication was confirmed by Wei and Lu [96], as they addressed that the regulation in terms of penalties may have a significant impact on the improvement of environmental quality. Jin et al. [35] also revealed that the heavy fine may restrain the environmentally unfriendly behavior of the enterprises.

5.3.6 *Summary*

This section applies evolutionary game theory combined with system dynamics to model the possible strategic actions of stakeholders (government, operating enterprise of the incineration plant, and residents) and predict their variation involved in the NIMBY conflicts with regard to waste incineration. The intervention of policy

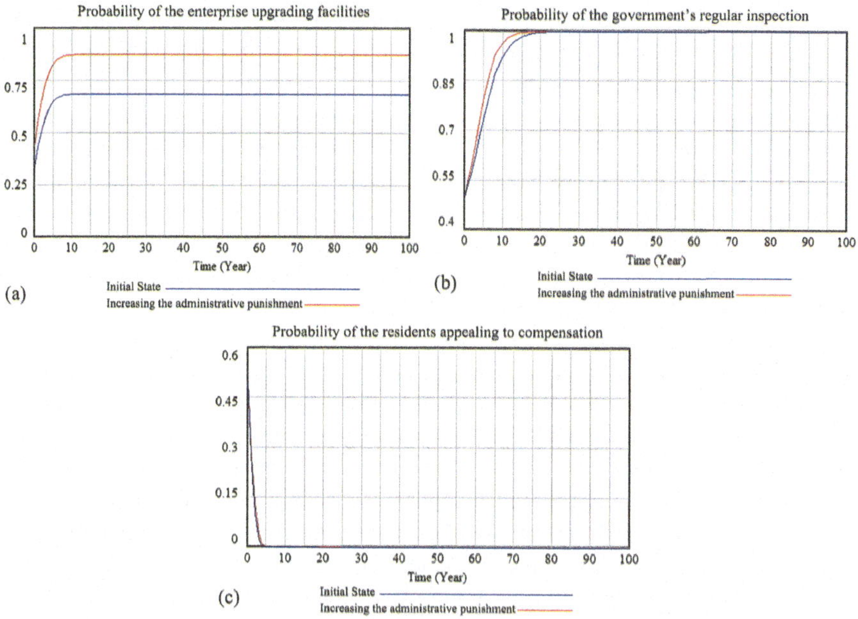

Fig. 5.16 Simulation results under change of administrative sanction. **a** Probability of the enterprise upgrading facilities. **b** Probability of the government's regular inspection. **c** Probability of the residents appealing to compensation

incentives and administrative punishments are incorporated into the game theoretical analysis, to discriminate their performance in mitigating the conflicts.

The game theoretical analysis shows that the three players with bounded rationality can reach equilibrium states without any external policy stimulus. For the strategic actions on the local government and residents, they will rapidly transform into a pair of dominant strategies throughout most of the simulation period, i.e. the government will implement regular inspection and residents will not complain about the project. No matter whether imposing policy incentives or administrative sanctions are imposed, their strategic actions do not have a change. Such results may imply that residents' action is not highly related to government policies involved in NIMBY conflicts. A possible reason may be that the government continues to conduct regular and strict inspection on the environmental management of the incineration plant to ensure the public interest, so the residents tend not to call for economic compensation [84].

The second implication is the strategic choice between incentives and punishments. The enterprise's decision making is closely relevant to government regulation. A number of empirical studies have found that the implementation of incentives is effective in promoting the enterprise's research and development , thus promoting technological innovation [110, 121]. However, the simulation result in the study shows that there is no significant change in the enterprise's action when government subsidies are provided. A possible reason could be that the enterprise itself, as a rational entity, is profit-oriented and takes the economic benefits as the bottom line of its operation [28]. In such a context, the enterprise gives high priority to maximizing its revenue to maintain its survival, while paying less attention to the corporate social responsibility. This is prominent in the industries of waste treatment and disposal in China, for which the government subsidy is an important revenue to cover the operation and maintenance costs [48, 108]. Without sufficient funding, the enterprises are not willing to make additional investments voluntarily to improve facilities to meet the NIMBY challenges [104].

The administrative penalty seems to have a greater impact on informing the enterprise to take active response to improve the environmental quality. It is apparent that the probability of the enterprise's action on upgrading facilities has been increased by 0.17, up to 0.83. The intensive punishment may increase the illegal cost to some extent, which has a deterrent effect on the enterprise's environmentally unfriendly behaviour [96]. The intensive punishment may accelerate the strict supervision of NIMBY project operation to ensure public interests and reduce the complaints [117]. However, the mechanism of sanction should be carefully designed to avoid trapping into a dilemma regarding the trade-off between economic development and environmental protection [57].

The third implication is that this study provides a way to resolve the conflicts by structuring a linear transformation to simplify the complex relationships among the three stakeholders. In essence, this game takes the enterprise's action as a bridge, not only to respond to government policy making, but also to link with the perceptions of the residents, shown in Fig. 5.17. The government intervention is important to drive the enterprise improve the quality of its operations environment via regular inspection and strict punishment. The residents may perceive the substantial changes in environmental quality with the enterprise's continuous efforts, thus reducing their risk perceptions and further decreasing possible actions on complaints or even claims for compensation.

5.4 Decision Support System for Hazardous Materials Transport

In order to manage accidents and any sudden environmental pollution resulting from hazardous materials (Hazmat) transportation, it is critical to build a Decision Support System (DSS) based on the information technology. More importantly, any DSS should reduce the potential environmental risk and aim to mitigate the likely impacts, such as minimizing casualties and property loss.

This Section describes a DSS for the transportation of hazardous materials by rail in China in order to reduce the serious consequences of sudden environmental pollution accidents. The management of hazardous materials transportation and

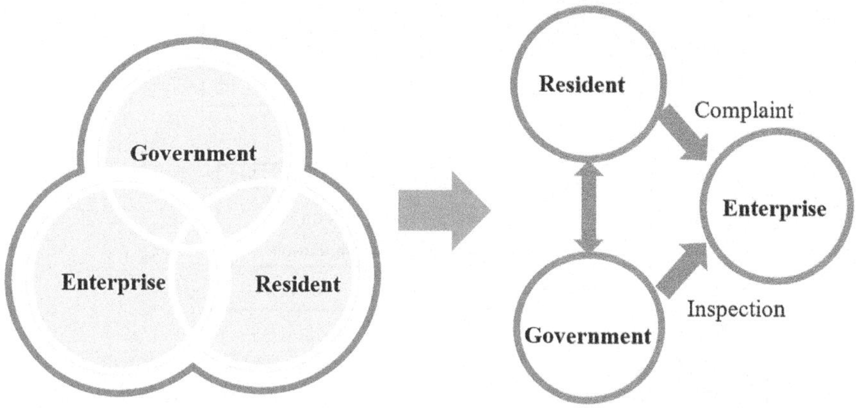

Fig. 5.17 The transformation of relationship among multi-player with governmental regulation

the resulting environmental risk are incorporated into the design approach for the system structure and functional framework. Moreover, management information system (MIS) and geographic information system (GIS) techniques are integrated to develop system functional modules within the DSS. This study mainly focuses on the data sheet construction in terms of MIS design, as well as the mathematical models applied to GIS buffer analysis. Upon completion of the system design and construction, selected functionalities of each sub-module are demonstrated. The limitations of the DSS are discussed, which lays a foundation for further development.

5.4.1 System Design

This decision support system design contains several elements, including environmental risk and safety, computer technology and surveying technology. On the one hand, the knowledge of environmental risk and safety can be transferred to the computer database to develop a basic MIS for hazardous materials transportation management. On the other hand, with the development of GIS techniques for DSS application, the environmental risk consequence analysis can more effectively identify the impacted area and provide a basis for evacuation. This interdisciplinary approach is an important contribution to this research area. Figure 5.18 illustrates the main stages in the development of the DSS.

In order to design the functional framework of the system, and determine the necessary attribute data for further database design, there is a need to identify the existing information regarding the management of hazardous materials throughout the rail system, especially the emergency response in the event of a sudden environmental accident, as shown in Fig. 5.19. When the emergency command centre at the Ministry of Railways receives an accident report, the exact location of the incident should be determined as soon as possible, and the necessary geographical information should

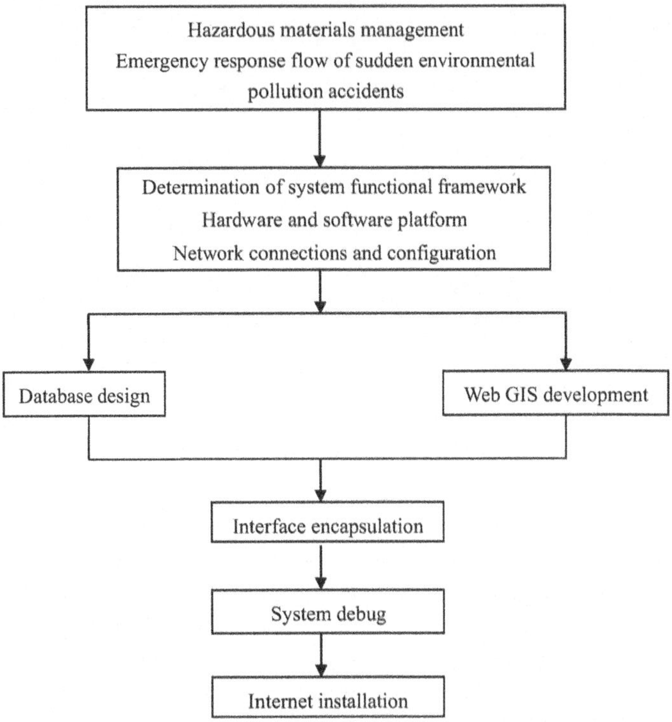

Fig. 5.18 Stages of the system design

be recorded, including the longitude and latitude, the distance between a residential area and the railway line, the nature of the local terrain, and so on. Meanwhile, such information as categories, amounts, and properties of hazardous materials for railway transportation can be retrieved from the Management Information System (MIS). Furthermore, the command centre designates the nearest environmental monitoring agency for local investigation. The agency can also query the MIS for the physical and chemical characteristics of the hazardous material in order to select the appropriate monitoring methods and equipment. Taking into account of the monitoring results, for the consequences of the accident can be assessed using the computer simulation, to predict the impacted area and the extent of pollution. Combined with expert advice, the local government can then take emergency measures immediately, such as evacuation of the local population, protection of waterways etc.

In the design of DSS for the Ministry of China Railways, the prerequisite is to integrate seamlessly with other existing official IT support systems, such as the Environmental Statistical Information System for the Ministry of China Railways (RESIS). RESIS has been developed by the China Academy of Railway Sciences to collect and transmit statistical data corresponding to various environmental indicators based on an information service platform, including the monitor index of air, water, waste, noise pollution, and so on. Moreover, the RESIS has an especially

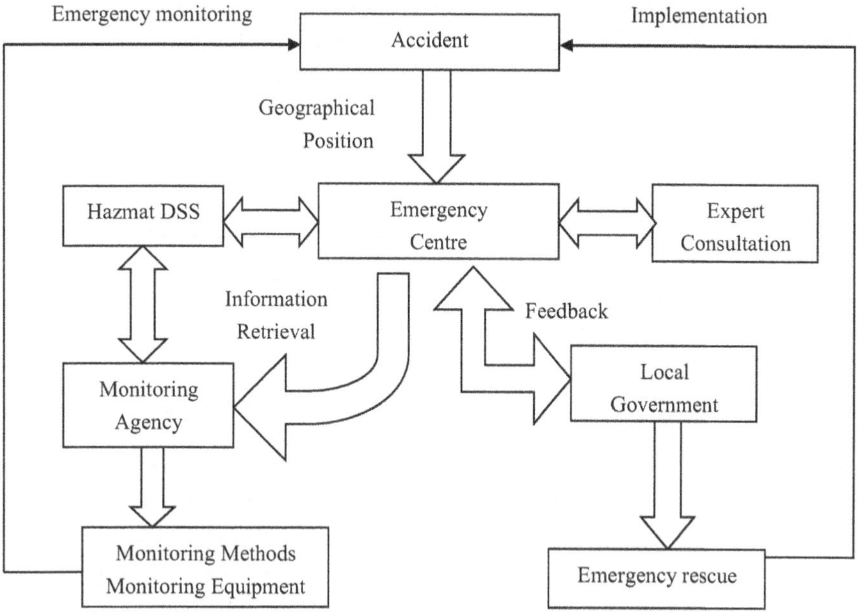

Fig. 5.19 Emergency response flow

pre-designed interface to connect with this proposed DSS for Hazardous Materials Railway Transportation, to facilitate information and resource sharing [25]. Thus, the hardware and software platforms for the development of the DSS, as well as the existing network connections and configuration, are required to maintain compatibility with RESIS. At the same time, the system functional framework needs to take into account the user requirements in the design of the different modules, thus supporting the database design. Additionally, the recent development of an internet-based GIS, i.e. Web GIS, is used to improve performance. Specifically, the Web GIS allows geographical information to be accessed via the internet [18]. Moreover, the proposed system database, which is described in more detail below (See Sect. 2.3), and the Web GIS are connected to provide a visual user interface. The final stages are the system debugging and internet installation. The system will also be tested for proper operation by inputting the necessary attribute data, such as transportation records of hazardous materials (hazmat), typical cases of sudden environmental pollution accidents, etc.

The system structure consists of a three-tier framework, including a user layer, a processing layer and a database layer (see Fig. 5.20). As different users log into the system, their instructions are first passed from the user layer to the next layer for processing. Subsequently, the processing layer accesses the database to retrieve, update, or revise the corresponding information. Finally, the available information is returned to the user.

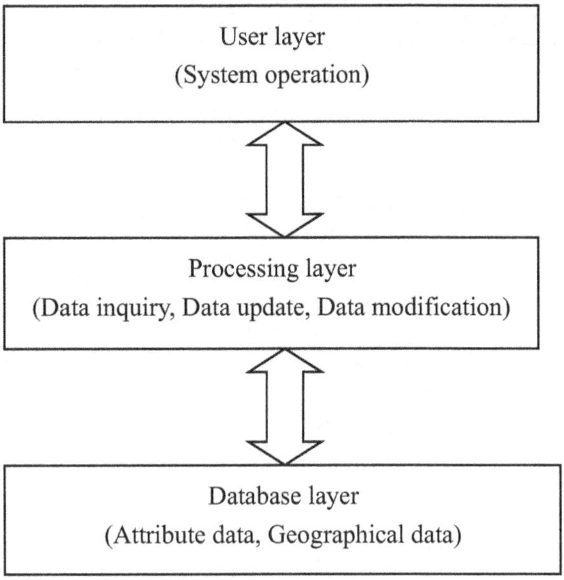

Fig. 5.20 Three Triple layer of the system structure

Figure 5.21 shows the system functional framework which divides the DSS into five main 'child' systems: hazmat transportation information, environmental security management, environmental simulation, environmental emergency monitoring and emergency response. In terms of the functionality of each subsystem, 'hazmat transportation information' and 'environmental security management' aim at the daily management of hazardous materials, with the former focusing on shipment tracking and the latter on information retrieval, such as physical and chemical properties of hazmat, human health risk, and typical environmental accident cases resulting from hazmat transportation. The 'Environmental simulation, emergency monitoring and response' pays more attention to the precaution and dealing with sudden environmental accidents, such as consequence analysis, environmental impact assessment, accident report, etc. Furthermore, all these subsystems are composed of correlative attribute modules, which is convenient for database design. The details of each attribute module are reflected in the following section. As the access to the system is restricted, the roles of users are mainly divided into administrators and normal users. Consequently, the authentication of the whole system is restricted to the administrators, including data searching, updating, deleting and modifying. Ordinary users can only access the system.

5.4.2 Database Design

As the MIS performance is affected directly by the database reliability and validity, the database design is a critical part of the system design, through which the

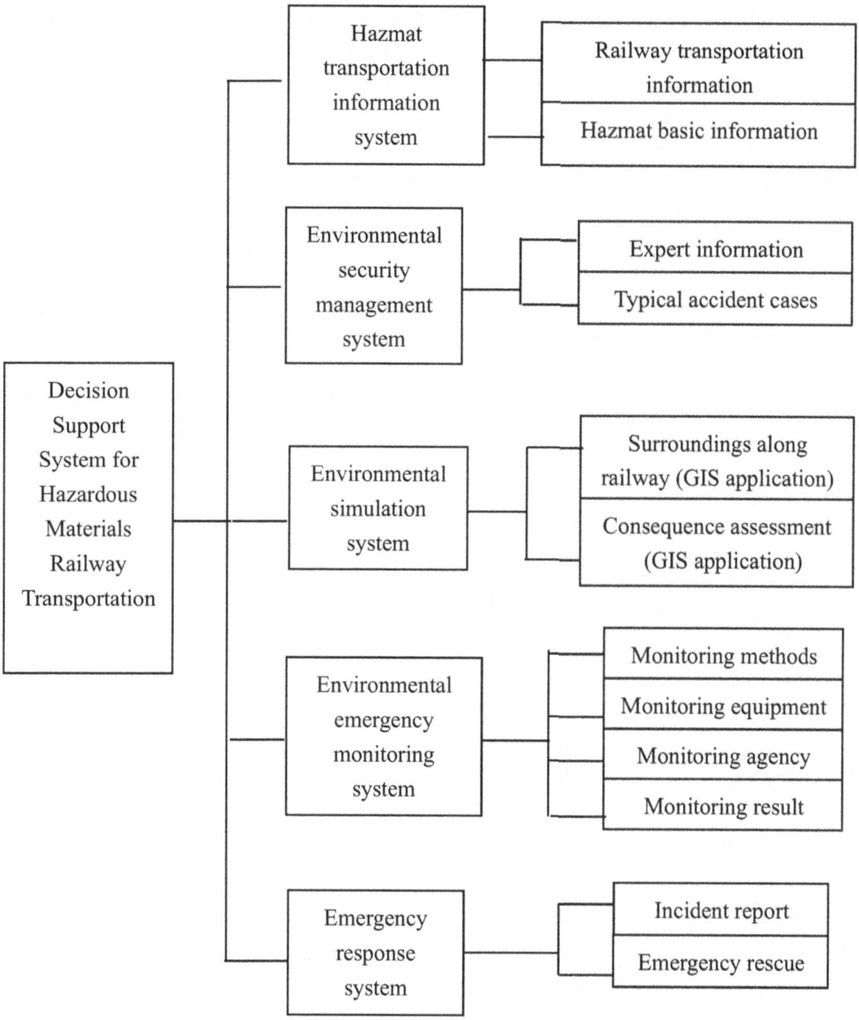

Fig. 5.21 System functional framework

information required by the DSS can be stored and retrieved. Since the existing office systems are based on ORACLE commercial database, it is used for the development of the database in this study [25]. Furthermore, Java Serve Page (JSP) is integrated with the Oracle database to operate the system securely by means of a web page.

The designed datasheets within each subsystem are shown below. Within the hazmat transportation information system, not only the shipment information can be tracked, but also the properties of hazardous materials can be retrieved. The attribute data of the transportation information module are selected from the relevant literature [38, 40], which include dangerous goods code, name, category, amounts, packaging method, loading technique, transaction stop, train code, both start and arrival stations, depository code, shipping company information, and so on. Dispatchers are required to inform the system operator of the local railway bureau so that the above information can be updated in the database in a timely manner, as soon as a real consignment is generated. A summary of the above attribute information is presented in Table 5.9. Moreover, the 'primary key' for the index is determined by merging the hazmat code, train code and depository code together. This detailed data is intended to be collected and updated by the local railway bureau, which is responsible for the shipment records of hazardous materials, to ensure the reliability of the data.

According to the existing list of hazardous materials which is published by "China Railway Ministry Dangerous Goods List of Railway Transportation" (2006),

Table 5.9 Hazmat transportation information

Field name	Data type	Instruction
Danger_code	Varchar2(4)	Hazardous goods code (PK)
Goods_name	Varchar2(20)	Goods name
Goods_charac	Varchar2(20)	Goods main character
Goods_type	Varchar2(20)	Goods category
Goods_num	Number(5,1)	Goods amounts
Num_unit	Varchar2(6)	Amount unit
Package_way	Varchar2(20)	Goods packaging way
Load_way	Varchar2(20)	Goods loading way
Goods_trans_stop	Varchar2(20)	Goods transaction stop
Train_no	Varchar2 (20)	Train code (PK)
Start	Varchar2(10)	Start station
Arrival	Varchar2(10)	Arrival station
Passing_station	Varchar2(500)	Main stops on the way
Store_cabin_no	Varchar2(10)	Depositary code (PK)
Store_cabin_type	Varchar2(20)	Depositary cabin type
Carriage_volume	Varchar2(15)	Carriage volume
Goods_source	Varchar2(100)	Goods source
Deliver_date	Date	Start time
Special_order	Varchar2(100)	Special regulation
Temp_store_infor	Varchar2(200)	Interim depositary
Shipping_name	Varchar2(50)	Shipping company name
Shipping_qualify	Varchar2(50)	Shipping qualification
Shipping_contact	Varchar2(100)	Shipping contact
Shipping_infor	Varchar2(200)	Shipping information

the 'Hazmat basic information' module will collect such information as summary of hazards, human injury and fatality, physical and chemical characteristics, and emergency treatment (see Table 5.10). Moreover, these data are routinely updated annually by the administrator of the China Railway Ministry, which is ultimately responsible for the logistics of hazardous materials.

In the subsystem of 'environmental security management', Table 5.11 shows the datasheet of the typical cases of environmental pollution accidents including case number, time and place of occurrence,, administrative region and station, etc., as well as attribute information of experts. When a sudden environmental pollution accident occurs, experts in the area closest to the site of the incident can be contacted for consultation as early as possible. This data can be accessed and updated annually by the safety administrator of the China Railway Ministry for statistical analysis of sudden environmental accidents caused by hazmat transportation.

According to the environmental monitoring results, it is urgent to implement safety protection and emergency rescue, thus to minimize negative consequences. The environmental emergency monitoring system consists of four functional modules including agency, equipment, methods and result. In addition, these modules contain hyperlinks for information retrieval. All the attribute information of an emergency monitoring system is presented in Table 5.12. The information is indexed by the same primary key, e.g. the monitoring agency code in these datasheets (see Table 5.12), to reduce data redundancy and thus increase query efficiency. This data should be provided and managed by the qualified agency of environmental monitoring attached to the local railway bureau.

The emergency response system is primarily used for the submission of incident reports. The Chinese government has categorized sudden environmental accidents into 4 classes (Central People's Government National contingency plan, 2006), as 'extraordinary serious accidents', 'major accidents', 'serious accidents', 'common frequent accidents', respectively. Regardless of the type of accident, the incident report should be immediately submitted to the railway command centre by the nearest stop or station for further decision-making. Meanwhile, the local government will carry out emergency treatment, such as communicating with the nearest hospital and fire station to implement rescue procedures, and wait for further instructions from the command centre (see Table 5.13).

5.4.3 Model for Environmental Consequence Assessment

GIS buffer analysis is one of the most important applications in this decision support system. Buffer analysis is used to identify or determine areas whether an area is inside or outside a specific area generated based upon the geographical features [95]. In this study, the buffer analysis is particularly used to determine whether the area surrounding the railway line is susceptible to a sudden environmental accident caused by the hazmat transportation. How to integrate the GIS with the MIS is the main challenge of this study due to the different development platforms. However,

Table 5.10 Basic information of hazmat

Field name	Data type	Instruction
Danger_No	Varchar2(10)	Hazardous goods code (PK)
CAS_no	Varchar2(20)	CAS number (summary)
Chi_Name	Varchar2(60)	Goods Chinese name (summary)
Eng_Name	Varchar2(150)	Goods English name (summary)
Danger_sign	Varchar2(20)	Dangerous sign (summary)
Main_use	Varchar2(500)	Main purpose (summary)
Pass	Varchar2(60)	Intruding way (harm and fatalness)
Hel_Harm	Varchar2(400)	Health harm (harm and fatalness)
Enviro _Affect	Varchar2(200)	Environment affect (harm and fatalness)
Fire_explode	Varchar2(200)	Fire and explode (harm and fatalness)
Acute_Noxious	Varchar2(200)	Acute toxicity (harm and fatalness)
Espe_Poison	Varchar2(200)	Special toxicity (harm and fatalness)
Thrill	Varchar2(200)	Thrill (harm and fatalness)
Chop	Varchar2(200)	Aberrance (harm and fatalness)
Cancer	Varchar2(200)	Carcinogenicity (harm and fatalness)
Abnormality	Varchar2(200)	Abnormality (harm and fatalness)
Byname	Varchar2(60)	Byname (physical and chemistry)
Mole_Formula	Varchar2(50)	Molecular formula (physical and chemistry)
Mole_Formula2	Varchar2(50)	Molecular weight (physical and chemistry)
Explode_Upper	Number(4,1)	Upper limit of exploding (physical and chemistry)
Explode_Lower	Number(4,1)	Lower limit of exploding (physical and chemistry)
Shape	Varchar2(100)	Appearance and shape (physical and chemistry)
Melt_Point	Varchar2(30)	Melting point (physical and chemistry)
Boil_Point	Varchar2(30)	Boiling point (physical and chemistry)
Density	Varchar2(30)	Comparative density (physical and chemistry)
Vapor_pressure	Varchar2(30)	Vapor pressure (physical and chemistry)
Flash_Point	Varchar2(30)	Flash point (physical and chemistry)
Dissolve	Varchar2(100)	Dissolvability (physical and chemistry)
Stability	Varchar2(10)	Stability (physical and chemistry)
Nature_Temper	Number(3,1)	Self-ignition temperature (physical and chemistry)
Critical_Temper	Number(3,1)	Critical temperature (physical and chemistry)
Leak_deal	Varchar2(2000)	Leakage treatment (emergent treatment)
Protect_way	Varchar2(2000)	Defending way (emergent treatment)
Emergency_way	Varchar2(2000)	First aid (emergent treatment)

Table 5.11 Environmental security management module

Field name	Data type	Instruction
Case_NO	Varchar2(4)	Case number (PK)
Occur_time	Date	Occurrence time
Occur_loc	Varchar2(50)	Occurrence locus
Admin	Varchar2(20)	Administrative region
Station	Varchar2(50)	Administrative station
Accident_Name	Varchar2(50)	Accident case name
Accident_Cause	Varchar2(600)	Accident causation
Expert_no	Varchar2(4)	Expert number (PK)
Name	Varchar2(10)	Name
Sex	Varchar2(4)	Gender
headship	Varchar2(50)	Headship
Professional_Title	Varchar2(50)	Title of a technical post
Office_tele	Varchar2(15)	Telephone
Working_address	Varchar2(50)	Work address
Professional_field	Varchar2(50)	Research field

the GIS application is encapsulated for consequence analysis as an ActiveX in the MIS functional module for the user to access.

In order to define the impacted area by using GIS for implementing emergency measures, it is necessary to model the environmental consequences resulting from events such as the diffusion of hazardous materials in the atmosphere, contamination of water supply, etc. Although meteorological information, such as local weather conditions, instantaneous wind speed etc. isa pre-requisite for accurate predictions, the proposed model assumes ideal conditions. In particular, to simplify the model calculation, only the initial wind direction at the time of gas leakage is considered in this study. As suggested by the China Technical Guidelines for Environmental Risk Assessment on Projects (National Environmental Protection Ministry HJ T169-2004), the multi-puff model of gas diffusion is recommended to estimate the concentration of a toxic gas. Importantly, it is assumed that the atmospheric effect is instantaneous. The multi-puff model can be described as follows:

$$C_w^i(x, y, 0, t_w) = \frac{2Q'}{(2\pi)^{3/2}\sigma_{x,eff}\sigma_{y,eff}\sigma_{z,eff}} exp(-\frac{H_e^2}{2\sigma_{x,eff}^2}) exp\left\{-\frac{(x-x_w^i)^2}{2\sigma_{x,eff}^2} - \frac{(y-y_w^i)^2}{2\sigma_{y,eff}^2}\right\} \tag{5.28}$$

where $C_w^i(x, y, 0, t_w)$ is the ground concentration of the of the i-th puff at the point coordinates $(x, y, 0)$ while the time of t_w. Q' is the release of the puff; where $Q' = Q\Delta t$; Q is the release rate, $(\text{mg}\cdot\text{s}^{-1})$; Δt is the release time, (s); $\sigma_{x,eff}$,$\sigma_{y,eff}$, $\sigma_{z,eff}$ are the equivalent diffusion parameters followed by x, y and z directions, respectively, which can be calculated by the following equations:

Table 5.12 Emergency monitoring module

Field name	Data type	Instruction
Monitor_orga_No	Varchar2(4)	Monitoring agency code (PK)
Monitor_orga_Name	Varchar2(20)	Monitoring agency name
Monitor_field	Varchar2(20)	Monitoring field
Orga_Addr	Varchar2(100)	Agency address
Orga_tele	Varchar2(15)	Agency telephone
Equip_No	Varchar2(4)	Equipment number
Equip_Name	Varchar2(20)	Equipment name
Equip_type	Varchar2(20)	Equipment type
Equip_Amount	Number(3)	Equipment amount
Amount_Unit	Varchar2(8)	Amount unit
Manufacture	Varchar2(50)	Manufactory
Measure_Range	Varchar2(50)	Measure range of equipment
Effect_Period	Varchar2(50)	Period of validity
Equip_Function	Varchar2(200)	Equipment function
Equip_deposit_Place	Varchar2(200)	Description of equipment depositary
Monitor_Way_No	Varchar2(4)	Monitoring methods code
Monitor_name	Varchar2(10)	Monitoring methods name
Monitor_element	Varchar2(300)	Monitoring factors
Monitor_description	Varchar2(600)	Description of monitoring methods
Monitor_Result	Varchar2(800)	Emergent monitoring result
Monitor_loc	Varchar2(50)	Emergent monitoring place

Table 5.13 Emergency response module

Field name	Data type	Instruction
Incident_code	Varchar2(4)	Incident code (PK)
Occur_time	Date	Occurring time
Occur_loc	Varchar2(50)	Occurring site
Hospital_No	Varchar2(4)	Hospital code (PK)
Hospital_Name	Varchar2(20)	Hospital name
Hospital_Addr	Varchar2(100)	Hospital address
Hospital_tele	Varchar2(15)	Hospital telephone
Hospital_Charge	Varchar2(10)	Principal name of hospital
Fire_station_No	Varchar2(4)	Fire station code (PK)
Fire_station _Name	Varchar2(20)	Fire station name
Fire_station _Addr	Varchar2(100)	Fire station address
Fire_station _tele	Varchar2(15)	Fire station telephone

$$\sigma_{j,eff}^2 = \sum_{k=1}^{w} \sigma_{j,k}^2 (j = x, y, z) \tag{5.29}$$

$$\sigma_{j,k}^2 = \sigma_{j,k}^2(t_k) - \sigma_{j,k}^2(t_{k-1}) \tag{5.30}$$

where x_w^i and y_w^i represent the barycentric coordinates of the ith puff at the end of time interval w. They can be solved separately by the following formulations.

$$x_w^i = u_{x,w}(t - t_{w-1}) + \sum_{k=1}^{w-1} u_{x,k}(t_k - t_{k-1}) \tag{5.31}$$

$$y_w^i = u_{y,w}(t - t_{w-1}) + \sum_{k=1}^{w-1} u_{y,k}(t_k - t_{k-1}) \tag{5.32}$$

The water dispersion model is a simplified one-dimensional water quality model that provides an estimate of the approximate length of a contaminated river. In this model, it is assumed that the liquid hazardous materials are dispersed instantaneously upon contact with the river, and thus ignored the mixing time. It is recommended in this study to use the instantaneous point source model to predict the environmental impact [31]. Without considering the boundary condition of the river, the ideal computational equation of the instantaneous point source model is presented below:

$$C(x, y, z, t) = \frac{Mexp[-(x - U_x t - x_0)^2/4E_x t]}{8\pi t(\pi E_x E_y E_z t)^{\frac{1}{2}}} exp(-kt) \times exp[-(y - y_0)^2/4E_y t] exp[-(z - z_0)^2/4E_x t] \tag{5.33}$$

where t is the time of leakage; M (mg) is the total amount of liquid hazardous material released,$M = C_0 V_0$; where C_0 (mg/L) is the concentration; $V_0(L)$ is the initial volume; k is the decay rate of pollutants, 1/s; x_0, y_0, z_0 are the point coordinates where the accident happened; U_x is the velocity of flow in the X direction, m/s; E_i is the dispersion coefficient, $(i = x, y, z)$, m^2/s, which can be estimated by the following equation.

$$E_i = a \cdot H \cdot u \tag{5.34}$$

where a is the dimensionless coefficient; H is the average depth of river, m; u is the frictional velocity, m/s, which can be roughly calculated as follows:

$$u = \sqrt{\frac{\tau}{\rho}} = \sqrt{gRS} \approx \sqrt{gHS} \tag{5.35}$$

where τ is the wall shear stress, kg/(m.s^2); ρ is the water density, kg/m^3; g is the acceleration of gravity, m/s^2; S is the hydraulic slope; R is the hydraulic radius, m.

Although these assumptions limit the application and accuracy of the model to some extent, they at least promise a rough assessment of the consequences to help the decision maker take emergency action to evacuate the local population and protect drinking water as soon as possible.

5.4.4 System Function

JSP is programmed to connect to the Oracle database and support interface operations. Some typical screenshots from each sub-module are selected to demonstrate the system operation in the following sections. As this system is built in Chinese for the benefit of the China Railway Ministry for further use, some English notes have added to the figures to demonstrate the system.

Figure 5.22 shows the login interface and control panel of the proposed DSS. Moreover, the system operation is restricted to use by the competent authorities only, so it is connected to the internal official network, and is not available for public use. The user roles are divided into two categories, as administrator and normal user, respectively (see Fig. 5.22a). In addition, the users are differentiated by their administrative roles, followed by the safety administrator of the Ministry of Railways, the local railway bureau and the base station or stop (see the login graph at Fig. 5.22b). All the operations on the system can be performed by the administrator of the Ministry of Railways, such as information upgrade, revision, maintenance, etc. Local railway bureau mainly focuses on system operation, such as daily transportation shipment tracking, coordination of emergency response, etc. However, normal users (base station) can only retrieve information from the system within the limits of their authority, except for submitting transportation safety and emergency reports. After a user logs in successfully by means of appropriate authentication, the web page is transferred to the user control panel automatically. The functional submodules are presented in the left side, and the login information, such as the user's login name, user rights, etc., are displayed on the right side. In addition, a quick search engine is provided to connect to the corresponding functional modules directly (see Fig. 5.22c). For instance, an environmental pollution accident case is displayed after selecting 'Typical environmental accident cases' from the drop-down list provided by the quick search engine. As an example, as shown in the Fig. 5.22d, a muriatic acid leakage caused by a train collision requires the evacuation of 400 people in the vicinity .

Figure 5.23 a shows selected screenshots from the subsystem of 'Hazmat Transportation Information'. The primary key presented in Table 5.9 is registered as a unique search index for a query. A short summary of the search index is displayed in the screenshot, which includes the hazmat code, name, train code, starts and arrival station. For instance, suppose the transportation administrator of Chengdu Railway Bureau wants to track the shipment information of sulfuric acid, upon submitting "Chengdu" as the start station to the system, the freight volume (2 tons), arrival

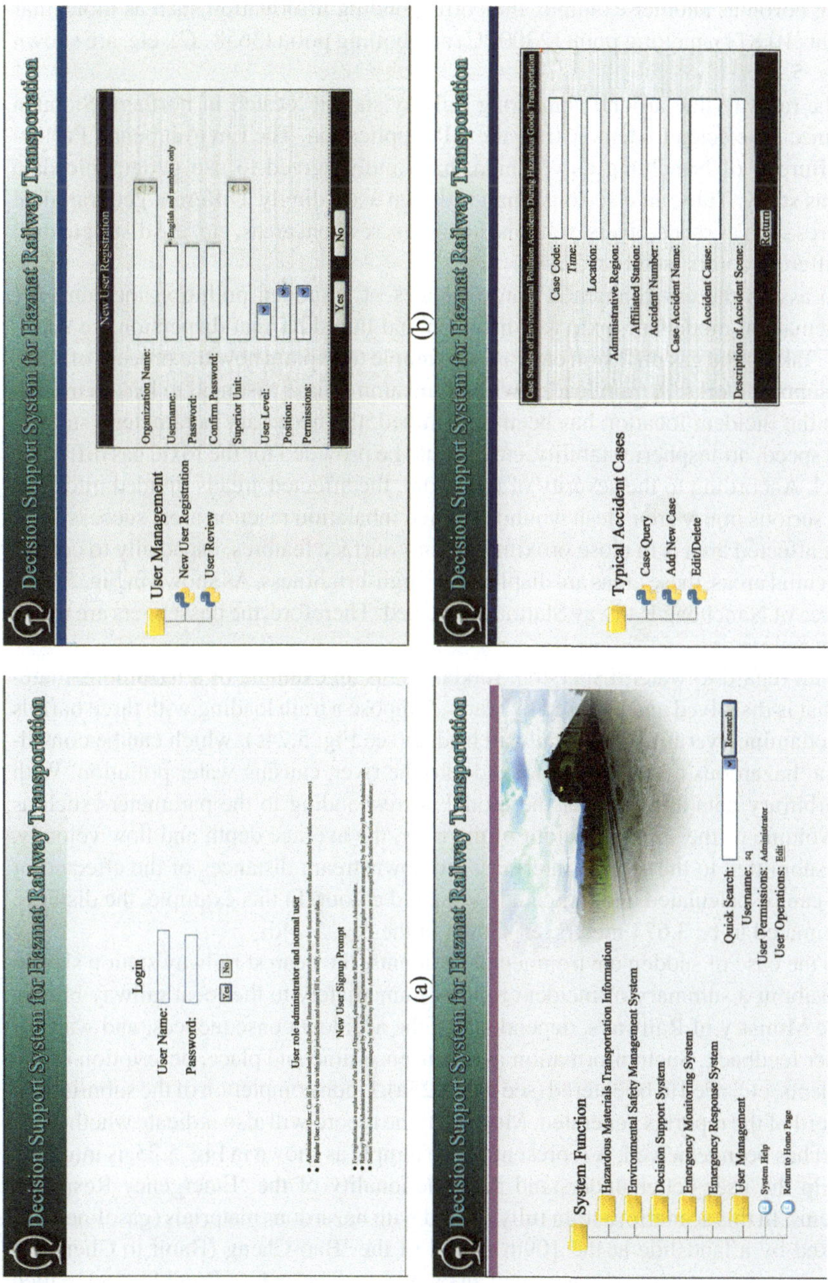

Fig. 5.22 System logon interface and control panel

station (Nanjing), etc. can be tracked immediately (see Fig. 5.23b). If the physical and chemical characteristics of a specific hazardous material need to be accessed, taking boron as another example, the corresponding information such as molecular weight (10.811), melting point (2300 °C) and boiling point (3658 °C), etc. are shown in Fig. 5.23c.

The railway line around Nanchong railway station located in northern Sichuan province is selected to demonstrate the GIS application. The Environmental Protection Bureau of Nanchong City, China, has kindly agreed to use geographic data for this study. Thus, an electronic map is drawn accordingly. Different geographical features such as roads, rivers, schools, hospitals, resident areas, etc. are distinguished by different colors (see Fig. 5.24a).

To assess the environmental consequences of a sudden pollution incident, the mathematical models of toxic gas diffusion and liquid hazmat dispersion are simulated. Taking the gas diffusion case as an example to explain how the process of decision support works. A train loaded with liquid ammonia is assumed to leak in transit. Once the incident location has been determined, the necessary parameters, such as wind speed, atmospheric stability, etc., should be provided for the toxic gas diffusion model. According to the severity of the injury, the affected area is divided into fatal area, serious injury area, flesh wound area and inhalation reaction area, successively. If the affected area is in close proximity to any surface features, especially to certain residential areas, these areas are displayed in high-brightness. As shown in Fig. 5.24b, the area of Nanchong Railway Station is affected. Therefore, the passengers are urged to evacuate.

With regard to water dispersion, rhodamine is an example of a hazardous material that is dissolved and diffused by water. Suppose a train loading with three barrels of rhodamine overturned on a railway bridge (see Fig. 5.24c), which can be considered a 'hazardous material' and leaked into the river, causing water pollution. With the arbitrary data assumed for the model corresponding to the parameters such as the volume of the spill, the width of the river, the average depth and flow velocity, etc., submitted to the system interface, the downstream distances of the affected of river can be calculated and indicated by the red colour. In this example, the distance is estimated to be 3,673 meters, as shown in the Fig. 5.24d.

In the case of sudden environmental accidents, the nearest railway station should also submit a summary of incident reports immediately to the local railway bureau or the Ministry of Railways, depending on the accident consequences, and wait for further feedback. Such information as occurrence time and place, description of the accidents, etc. need to be entered (see Fig. 5.25a). Upon completion of the submission, a record of the report is generated. Moreover, the record will also indicate whether the report has been reviewed. A representative example as shown in Fig. 5.25, is intended to help the user better understand the functionality of the 'Emergency Response System'. In this scenario, a train fully loaded with hazardous materials (gasoline) was blocked by a landslide at the 109th tunnel of the 'Bao-Cheng (Baoji to Chengdu) Railway Line', causing a leakage and an explosion. The staff of Baoji Station, which is the nearest base station, should be the first to arrive at the accident site and should

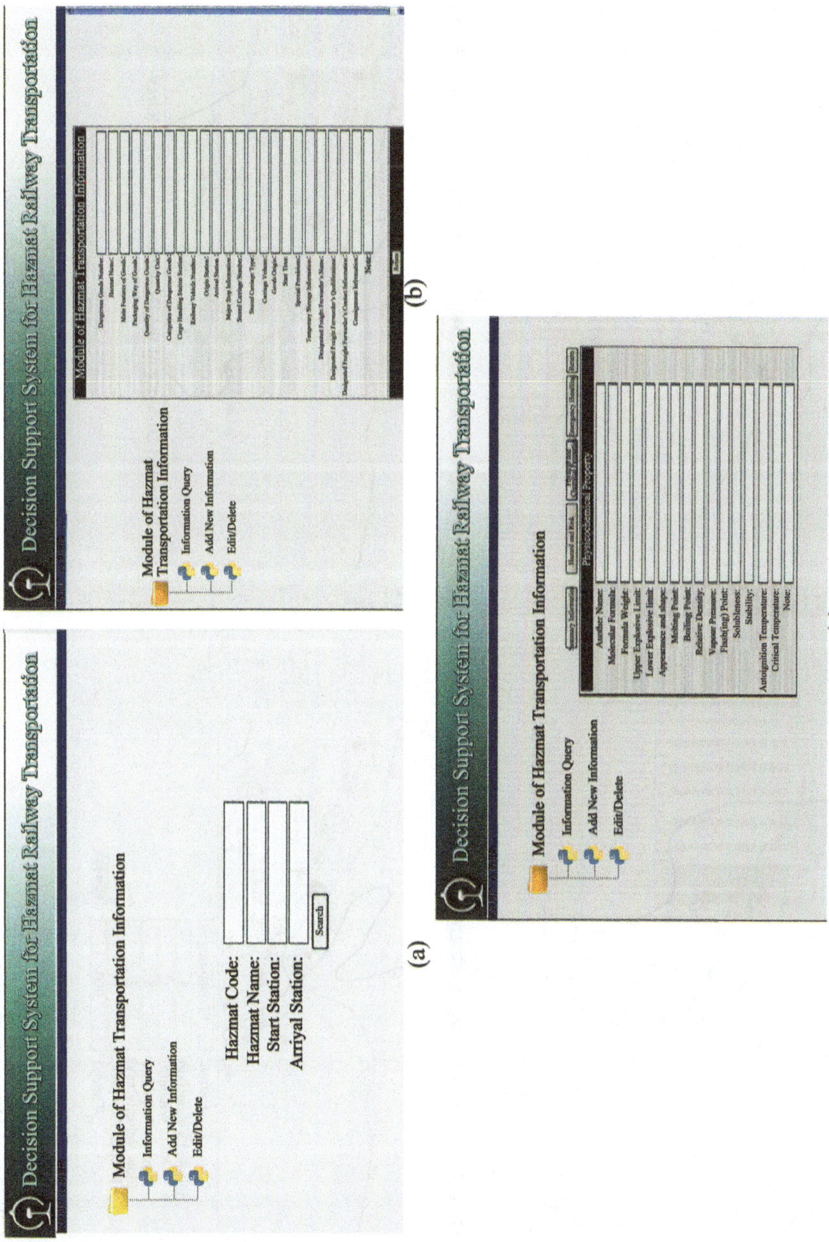

Fig. 5.23 Hazmat transportation information system

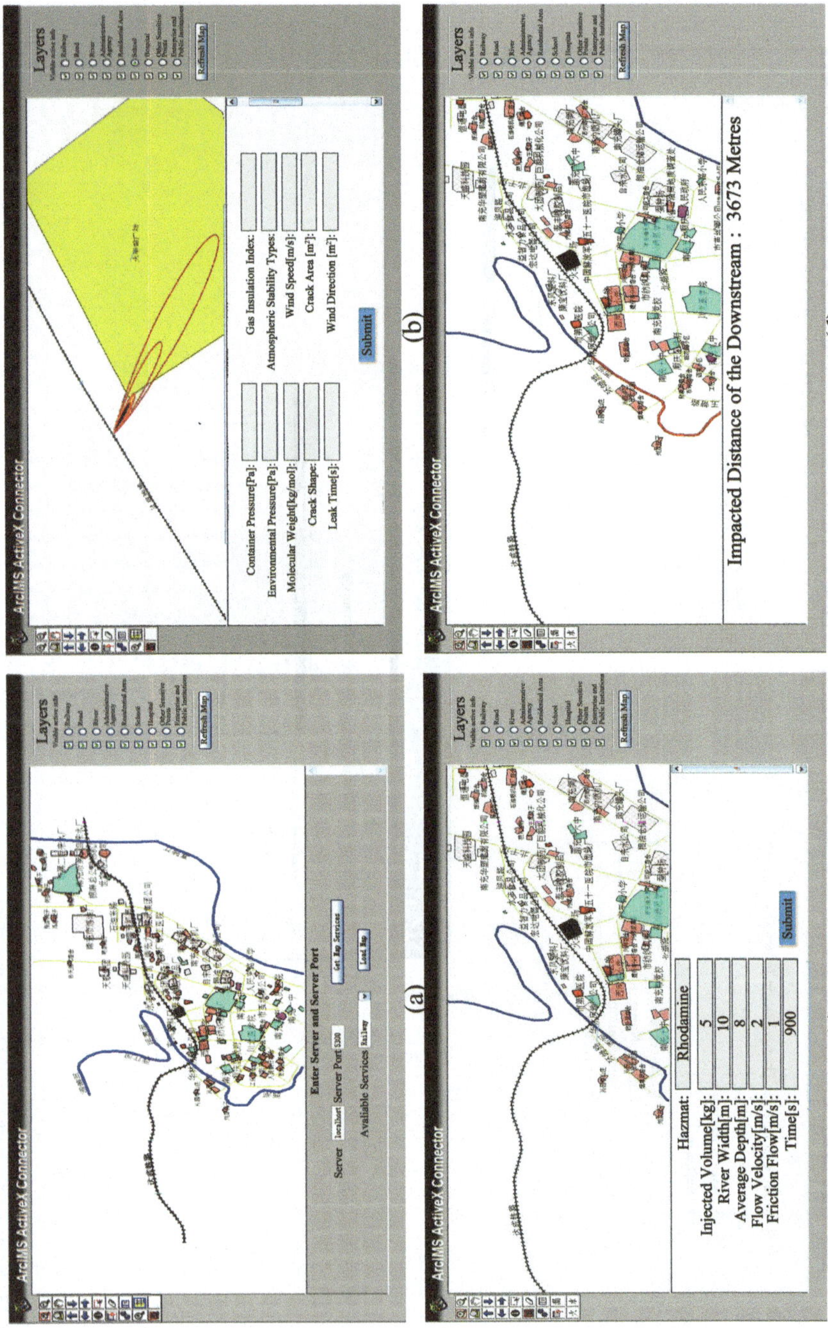

Fig. 5.24 GIS application in environmental simulation system

submit an initial incident report to the command center of the Ministry of Railways, and wait for the emergency rescue authorities to respond.

Within the subsystem of environmental emergency monitoring, the module of emergency monitoring equipment has been selected to demonstrate its function. For instance, if the soil or hydrological condition needs to be investigated in terms of the emergency monitoring requirement, the item of 'geological detector' can be accessed by the module to help choose the appropriate equipment (see Fig. 5.25c). In addition, the environmental monitoring agency affiliated with the local railway bureau is required to update such information as environmental monitoring methods, equipment annually, and to post the emergency monitoring results of a pollution incident, which are shown in Fig. 5.25d.

5.4.5 Summary

This section provides a design process of the decision support system DSS (DSS) for rail transportation of hazardous materials in China, including such activities as design of the system structure and functional framework, database sheet construction for MIS, mathematical model of environmental impact analysis for GIS. An interactive and user-friendly system interface is supported by JSP web page to connect with Oracle database to help users operate all the functional modules of the system. System operations such as information retrieval, GIS application for environmental impact prediction, accident report submission, etc. are demonstrated by selecting screenshots from each subsystem. Although the computer techniques for the system development are complex, the system functional framework is easy for users to assimilate and expand. Moreover, the modular design of the DSS uses standard programming languages and methodologies, to facilitate system maintenance and expand its application. The proposed system will be further improved and integrated with other existing official IT systems supporting the Ministry of China Railways to illustrate its advantages in the transportation management and sudden environmental pollution pre-warning.

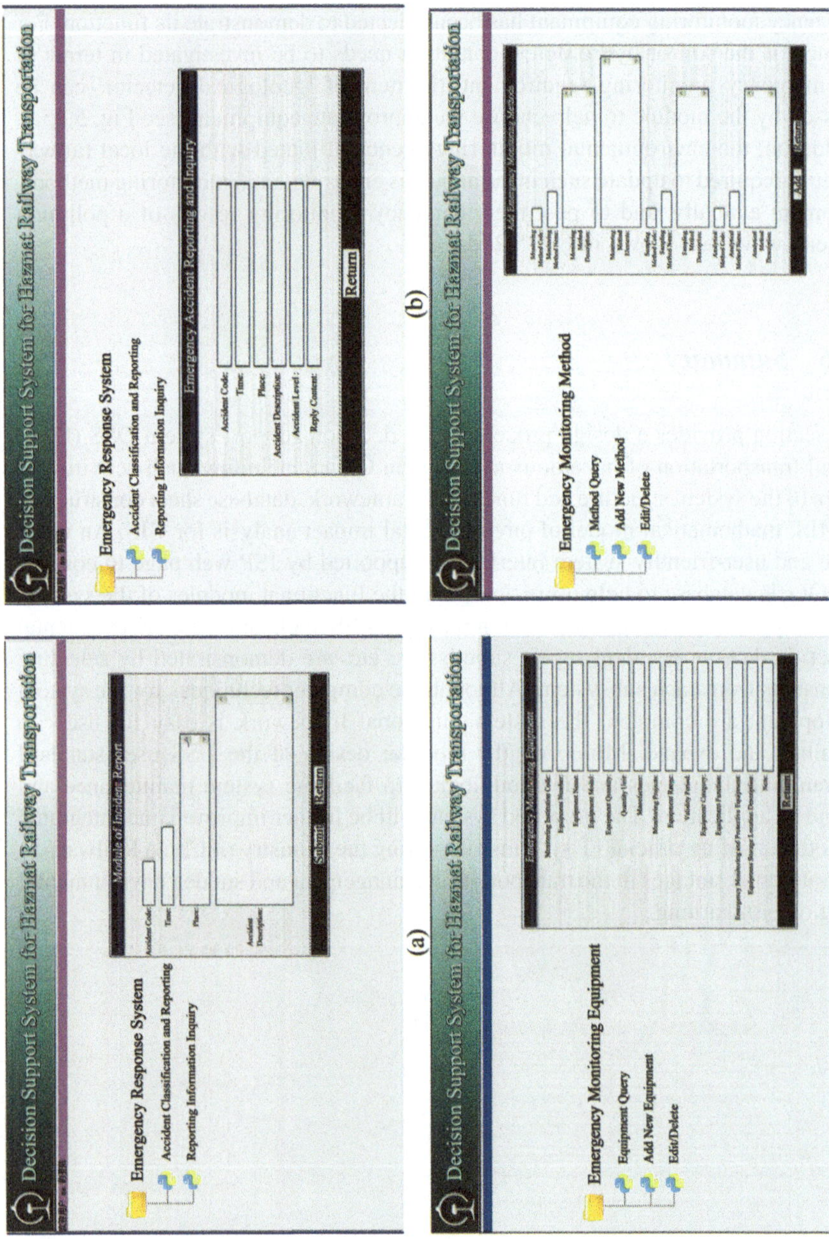

Fig. 5.25 Environmental monitoring and emergency system

References

1. Ahmed FN, Lan CQ (2012) Treatment of landfill leachate using membrane bioreactors: a review. Desalination 287:41–54
2. Bagchi A (2004) Design of landfills and integrated solid waste management, 3rd edn. Wiley, New York, NY
3. Beyene HD, Werkneh AA, Ambaye TG (2018) Current updates on waste to energy (WtE) technologies: a review. Renewable Energy Focus 24:1–11
4. Bian YT, Xu L, Li JS, Liu XQ (2016) Dynamical evolution of trading behavior on anti-coordination game in complex networks. China Finance Rev Int 6(4):367–439
5. Bicheldey TK, Latushkina EN (2010) Biogass emission prognosis at the landfills. Int J Environ Sci Technol 7(4):623–628
6. Bilgili MS, Demir A, Varank G (2009) Evaluation and modeling of biochemical methane potential (BMP) of landfilled solid waste: a pilot scale study. Biores Technol 100(21):4976–4980
7. Bosmans A, Vanderreydt I, Geysen D, Helsen L (2013) The crucial role of waste-to-energy technologies in enhanced landfill mining: a technology review. J Clean Prod 55:10–23
8. Broun R, Sattler M (2016) A comparison of greenhouse gas emissions and potential electricity recovery from conventional and bioreactor landfills. J Clean Prod 112:2664–2673
9. Burnley S, Coleman T, Peirce A (2015) Factors influencing the life cycle burdens of the recovery of energy from residual municipal waste. Waste Manage 39:295–304
10. Cakir AK, Gunerhan H, Hepbasli A (2016) A comparative study on estimating the landfill gas potential: modeling and analysis. Energy SourcesPart A Recov Utiliz Environ Effects 38(16):2478–2486
11. Cheng H, Zhang Y, Meng A, Li Q (2007) Municipal solid waste fueled power generation in China: a case study of waste-to-energy in Changchun City. Environ Sci Technol 41(21):7509–7515
12. Chhipi-Shrestha G, Rodriguez M, Sadiq R (2019) Selection of sustainable municipal water reuse applications by multi-stakeholders using game theory. Sci Total Environ 650:2512–2526
13. Christensen TH, Cossu R, Stegmann R (1992) Landfilling of waste: leachate. Elsevier Applied Science, London, pp 3–5
14. Cocarta DM, Rada EC, Ragazzi M, Badea A, Apostol T (2009) A contribution for a correct vision of health impact from municipal solid waste treatments. Environ Technol 30(9):963–968
15. Da Silva LJDB, Dos Santos IFS, Mensah JHR, Gonalves ATT, Barros RM (2020) Incineration of municipal solid waste in Brazil: an analysis of the economically viable energy potential. Renew Energy Int J 149:1386–1394
16. De Feo G, De Gisi S, Williams ID (2013) Public perception of odour and environmental pollution attributed to MSW treatment and disposal facilities: a case study. Waste Manage 33(4):974–987
17. DEFRA (Department for Environment, Food and Rural Affairs). (2007). Waste strategy for England. http://www.defra.gov.uk/environment/waste/strategy/strategy07/index.htm
18. Dhiman R, VishnuRadhan R, Inamdar AB, Eldho I (2020) Web-GIS integrated open source mashup technology as a cue for integrated management in coastal megacities. J Coast Conserv 24(2):18
19. Ding G (2011) Study of business environment and strategy of Qinhuangdao garbage power plant. Master thesis, North China Electric Power University, Beijing, China
20. El-Haggar S (2007) Sustainable industrial design and waste management: cradle to cradle for sustainable development. Elsevier Academic Press, US
21. Fan R, Dong L (2018) The dynamic analysis and simulation of government subsidy strategies in low-carbon diffusion considering the behavior of heterogeneous agents. Energy Policy 117(1):252–262

22. Fernández-Nava Y, Del Rio J, Rodríguez-Iglesias J, Castrillón L, Marañón E (2014) Life cycle assessment of different municipal solid waste management options: a case study of Asturias (Spain). J Clean Prod 81:178–189
23. Friedman D (1991) Evolutionary game in economics. Econometrica 59(3):637–666
24. Fu Y, Qiu Z, Fu C, Zhu D, Yin Y (2015) Analysis of municipal solid waste composition and physicochemical properties in urban areas of Chengdu. Sichuan Environ 33:126–129 (in Chinese with English abstract)
25. Guo HY, Zong Y, Liu ZD, Jia XF (2007) Environmental protecting guide line research of railway transport enterprises. Railway Occup Safety Health Environ Protect 34(5):235–237 (in Chinese)
26. Halkos GE, Papageorgiou GJ (2018) Pollution, environmental taxes and public debt: a game theory setup. Econ Anal Policy 58:111–120
27. Hettiaratchi P, Jayasinghe P, Tay JH, Yadav S (2015) Recent advances of biomass waste to gas using landfill bioreactor technology—a review. Curr Org Chem 19(5):413–422
28. Hjaila K, Puigjaner L, Laínez JM, Espuña A (2017) Integrated game-theory modelling for multi enterprise-wide coordination and collaboration under uncertain competitive environment. Comput Chem Eng 98:209–235
29. Ho WS, Hashim H, Lim JS, Lee CT, Sam KC, Tan ST (2017) Waste management pinch analysis (WAMPA): application of pinch analysis for greenhouse gas (GHG) emission reduction in municipal solid waste management. Appl Energy 185:1481–1489
30. Hogland W, Marques M, Nimmermark S (2004) Landfill mining and waste characterization: a strategy for remediation of contaminated areas. J Mater Cycles Waste Manage 6:119–124
31. Hu EB (2000) Environmental risk assessment: practical technique and methods. China Environmental Science Press, Beijing (in Chinese)
32. Huang Y, Ning Y, Zhang T, Fei Y (2015) Public acceptance of waste incineration power plants in China: Comparative case studies. Habitat Int 47:11–19
33. Ionescu G, Rada EC, Cioca LI (2015) Municipal solid waste sorting and treatment schemes for the maximization of material and energy recovery in a latest EU member. Environ Eng Manag J 14(11):2537–2544
34. Jha AK, Singh SK, Singh GP, Gupta PK (2011) Sustainable municipal solid waste management in low income group of cities: a review. Trop Ecol 52(1):123–131
35. Jin Y, Cheng C, Zeng H (2020) Is evil rewarded with evil? The market penalty effect of corporate environmentally irresponsible events. Bus Strateg Environ 29(3):846–871
36. Johari A, Ahmed SI, Hashim H, Alkali H, Ramli M (2012) Economic and environmental benefits of landfill gas from municipal solid waste in Malaysia. Renew Sustain Energy Rev 16:2907–2912
37. Johnson T (2013) The politics of waste incineration in Beijing: the limits of a top-down approach? J Environ Planning Policy Manage 15(1):109–128
38. Kang ZQ, Li LQ (2001) Computer application in dangerous goods transportation management. Railway Freight Transp 2:28–30 (in Chinese with English abstract)
39. Lang G, Xu Y (2013) Anti-incinerator campaigns and the evolution of protest politics in China. Environ Polit 22(5):832–848
40. Lei DY, Wen HC (2004) Computer inspecting and managing system of railway dangerous goods transportation. J Traff Transp Eng 4(2):124–125
41. Li W, Zhou Q, Hua T (2010) Removal of organic matter from landfill leachate by advanced oxidation processes: a review. Int J Chem Eng. Article ID 270532
42. Li Y, Zhao X, Li Y, Li X (2015) Waste incineration industry and development policies in China. Waste Manage 46:234–241
43. Li Z, Shen GQ, Alshawi M (2014) Measuring the impact of prefabrication on construction waste reduction: an empirical study in China. Resour Conserv Recycl 91:27–39
44. Liu LX, Zhu YC, Guo SB (2020) The evolutionary game analysis of multiple stakeholders in the low-carbon agricultural innovation diffusion. Complexity. Article ID 6309545
45. Liu M, Shadbegian R, Zhang B (2017) Does environmental regulation affect labor demand in China? Evidence from the textile printing and dyeing industry. J Environ Econ Manag 86:277–294

46. Liu Y, Ge YJ, Xia B, Cui CY, Jiang XY, Skitmore M (2019) Enhancing public acceptance towards waste-to-energy incineration projects: Lessons learned from a case study in China. Sustain Cities Soc 48:101582
47. Lombardi L, Carnevale E, Corti A (2015) A review of technologies and performances of thermal treatment systems for energy recovery from waste. Waste Manage 37:26–44
48. Lyamu HO, Anda M, Ho G (2020) A review of municipal solid waste management in the BRIC and high-income countries: a thematic framework for low-income countries. Habitat Int 95:102097
49. Ma L, Zhang L (2020) Evolutionary game analysis of construction waste recycling management in China. Resour Conserv Recycl 161:104863
50. Marshall RE, Farahbakhsh K (2013) Systems approaches to integrate solid waste management in developing countries. Waste Manage 33(4):988–1003
51. Martin-Utrillas M, Reyes-Medina M, Curiel-Esparza J, Canto-Perello J (2015) Hybrid method for selection of the optimal process of leachate treatment in waste treatment and valorization plants or landfills. Clean Technol Environ Policy 17(4):873–885
52. MEEC (Ministry of Ecology and Environment of the People's Republic of China) (2014) Standard for pollution control on the municipal solid waste incineration. http://www.mee.gov.cn/ywgz/fgbz/bz/bzwb/gthw/gtfwwrkzbz/201405/t20140530_276307.shtml (in Chinese)
53. MEEC (Ministry of Ecology and Environment of the People's Republic of China) (2018) Measures on public participation in environmental impact assessment. http://www.mee.gov.cn/gkml/sthjbgw/sthjbl/201808/t20180803_447662.htm (in Chinese)
54. MFC (Ministry of Finance of the People's Republic of China) (2013) Administrative measures on travel expenses of central and state organs. http://www.mof.gov.cn/mofhome/shenzhen/lanmudaohang/zhengcefagui/201408/t20140812_1125718.html (in Chinese)
55. Mian MM, Zeng XL, Bin Nasry AA, Al-Hamadani SM (2016) Municipal solid waste management in China: a comparative analysis. J Mater Cycles Waste Manage 19(3):1127–1135
56. Ministry of Commerce People's Republic of China (2007) Law of the People's Republic of China on prevention and control of environmental pollution by solid waste. http://english.mofcom.gov.cn/aarticle/policyrelease/internationalpolicy/200703/20070304471567.html.
57. Mira C, Thaleia K, Andy VDD, Jens S (2019) Trade-off between the economic and environmental impact of different decarbonisation strategies for residential buildings. Build Environ 155:137–144
58. MOHURD (Ministry of Housing and Urban-Rural Development of the People's Republic of China) (2009) Technical specification for operation maintenance and safety of municipal solid waste incineration plant (CJJ128-2009). China Architecture & Building Press (in Chinese)
59. Nair A, Sartaj M, Kennedy K, Coelho NM (2014) Enhancing biogas production from anaerobic biodegradation of the organic fraction of municipal solid waste through leachate blending and recirculation. Waste Manage Res 32(10):939–946
60. NDRC (National Development and Reform Commission) (2006) Trial Measures for the management of The Price and Cost-sharing of renewable energy power generation. http://www.gov.cn/ztzl/2006-01/20/content_165910.htm (in Chinese)
61. Nguyen KLP, Chuang YH, Chen HW, Chang CC (2020) Impacts of socioeconomic changes on municipal solid waste characteristics in Taiwan. Resour Conserv Recycl 163:105024
62. Niskanen A, Värri H, Havukainen J, Uusitalo V (2013) Enhancing landfill gas recovery. J Clean Prod 55:67–71
63. NPC (The 13th National People's Congress of the People's Republic of China) (2018) Law of the People's Republic of China on prevention and control of air pollution. http://www.mzyfz.com/html/2015/2019-01-03/content-1379287.html (in Chinese)
64. Peavy HS, Rowe DR, Tchobanoglous G (1985) Environmental engineering. McGraw-Hill, New York, NY, pp 631–633
65. Pires A, Martinho G, Chang NB (2011) Solid waste management in European countries: a review of systems analysis techniques. J Environ Manage 92(4):1033–1050
66. Prosperi M, Lombardi M, Spada A (2019) Ex ante assessment of social acceptance of small-scale agro-energy system: a case study in southern Italy. Energy Policy 124:346–354

67. Qian LY, Xu ZM, Pan HF, Pan SP (2016) Reachability analysis of new standard for domestic incinerator flue gas. Admin Tech Environ Monitor 28(2):57–60 (in Chinese with English abstract)
68. Ren X, Che Y, Yang K, Tao Y (2016) Risk perception and public acceptance toward a highly protested waste-to-energy facility. Waste Manage 48:528–539
69. Renou S, Givaudan JG, Poulain S, Dirassouyan F, Moulin P (2008) Landfill leachate treatment: review and opportunity. J Haz Mater 150:468-493
70. Sheng JC, Zhou WH, Zhu BZ (2020) The coordination of stakeholder interests in environmental regulation: lessons from China's environmental regulation policies from the perspective of the evolutionary game theory. J Clean Prod 249:119385
71. Smith JM (1982) Evolution and the theory of games. Cambridge University Press, Cambridge
72. Smith RL, Sengupta D, Takkellapati S, Lee CC (2015) An industrial ecology approach to municipal solid waste management: II. Case studies for recovering energy from the organic fraction of MSW. Resour Conserv Recycl 104:317–326
73. Smith RL, Sengupta D, Takkellapati S, Lee CC (2015) An industrial ecology approach to municipal solid waste management: I. Methodology. Resour Conserv Recycl 104:311–316
74. Song JB, Song DR, Zhang DL (2014) Modeling the concession period and subsidy for BOT waste-to-energy incineration projects. J Constr Eng Manage 141(10):4015033-1–4015033-12
75. SPBS (Sichuan Provincial Bureau of Statistics) (2019) An announcement on the average salary of all employees in cities and towns across the province in 2019. http://tjj.sc.gov.cn/sctjj/001/2020/5/22/eefbba86177a4210985af68d546b3b5d.shtml (in Chinese)
76. Subiza-Perez M, Marina LS, Irizar A, Gallastegi M, Anabitarte A, Urbieta N, Babarro I, Molinuevo A, Vozmediano L, Ibarluzea J (2020) Explaining social acceptance of a municipal waste incineration plant through sosciodemographic and psycho-environmental variables. Environ Pollut 263:114504
77. Sun C, Ouyang X, Meng X (2019) Public acceptance towards waste-to-energy power plants: a new quantified assessment based on 'willingness to pay.' J Environ Planning Manage 62(14):2459–2477
78. Sun LL, Yung EHK, Chan EHW, Zhu DJ (2016) Issues of NIMBY conflict management from the perspective of stakeholders: a case study in Shanghai. Habitat Int 53:133–141
79. Talyan V, Dahiya R, Anand S, Sreekrishnan T (2007) Quantification of methane emission from municipal solid waste disposal in Delhi. Resour Conserv Recycl 50:240–259
80. Tchbanoglous G, Kreith F (2002) Handbook of solid waste management, 2nd edn. The McGraw-Hill Companies, New York
81. Tong YQ, Liu JF, Liu SZ (2019) China is implementing 'Garbage Classification' action. Environ Pollut 259:113707
82. Urtiaga A, Rueda A, Anglada A, Ortiz I (2009) Integrated treatment of landfil leachates including electrooxidation at pilot plant scale. J Hazard Mater 166:1530–1534
83. Voytenko Y, McCormick K, Evans J, Schliwa G (2016) Urban living labs for sustainability and low carbon cities in Europe: towards a research agenda. J Clean Prod 123:45–54
84. Wan C, Shen GQ, Choi S Differential public support for waste management policy: the case of Hong Kong. J Clean Prod 175:477–488
85. Wang G, Deng J, Ma Z, Hao J, Jiang J (2018) Characteristics of filterable and condensable particulate matter emitted from two waste incineration power plants in China. Sci Total Environ 639:695–704
86. Wang J, Li Z, Tam VWY (2015) Identifying best design strategies for construction waste minimization. J Clean Prod 92:237–247
87. Wang J, Li Z, Wang X (2012) A system dynamic model for cost-benefit analysis of construction waste. Constr Econ 8:95–98 (in Chinese with English abstract)
88. Wang T, Lin B (2014) China's natural gas consumption and subsidies—from a sector perspective. Energy Policy 65:541–551
89. Wang W, Han F, Yuan G, Jiang J, Wang Z (2001) Study on methane generation from landfills in China. China Biogas 19(2):20–24 (in Chinese with English abstract)

90. Wang X, Jia M, Chen X, Xu Y, Lin X, Kao CM, Chen S (2014) Greenhouse gas emissions from landfill leachate treatment plants: a comparison of young and aged landfill. Waste Manage 34:1156–1164
91. Wang X, Jia M, Zhang C, Chen S, Cai Z (2017) Leachate treatment in landfills is a significant N_2O source. Sci Total Environ 596:18–25
92. Wang Y, Geng S, Zhao P, Du H, He Y, Crittenden J (2016) Cost–benefit analysis of GHG emission reduction in waste to energy projects of China under clean development mechanism. Resour Conserv Recycl 109:90–95
93. Wang H, Wang YN, Li X, Sun Y, Wu H, Chen D (2016) Removal of humic substances from reverse osmosis (RO) and nanofiltration (NF) concentrated leachate using continuously ozone generation-reaction treatment equipment. Waste Manage 56:271–279
94. Wang Z, Wang QX, Chen BX, Wang Y (2020) Evolutionary game analysis on behavioral strategies of multiple stakeholders in E-waste recycling industry. Resour Conserv Recycl 155:104618
95. Wang ZG, Du XSH (2016) Monitoring Natural World Heritage Sites: optimization of the monitoring system in Bogda with GIS-based multi-criteria decision analysis. Environ Monit Assess 188(7):384
96. Wei J, Lu S (2015) Investigation and penalty on major industrial accidents in China: the influence of environmental pressures. Saf Sci 76:32–41
97. Weibull JW (1997) Evolutionary game theory. MIT Press, Cambridge, Mass
98. Williams PT (2005) Waste treatment and disposal, 2nd edn. Wiley, Chichester, pp 220–227
99. Woon KS, Lo IMC (2013) Greenhouse gas accounting of the proposed landfill extension and advanced incineration facility for municipal solid waste management in Hong Kong. Sci Total Environ 458:499–507
100. Wu Z, Zheng D (2012) Generation regularities and application of LFG from Chengdu Chang'an waste landfill site. Environ Sanit Eng 20:46-48 (in Chinese)
101. Wu LH, Ma TS, Bian YC, Li SJ, Yi ZQ (2020) Improvement of regional environmental quality: Government environmental governance and public participation. Sci Total Environ 717:137265
102. Xin GZ, Gui WJ, Ang L, Yun L (2016) Technology, cost, a performance of waste-to-energy incineration industry in China. Renew Sustain Energy Rev 55:115–130
103. Xu JP, Zhu MY, Zhao SW (2020) Leader-follower optimized approach for carbon-economy equilibrium in the municipal solid waste (MSW) incineration industry. Environ Sci Pollut Res 27(26):32637–32658
104. Xu Y, Chan AP, Xia B, Qian QK, Liu Y, Peng Y (2015) Critical risk factors affecting the implementation of PPP waste-to-energy projects in China. Appl Energy 158:403–411
105. Xu YP, Zhou YQ, Wang DH, Chen SH, Liu JX, Wang ZJ (2008) Occurrence and removal of organic micropollutants in the treatment of landfill leachate by combined anaerobic-membrane bioreactor technology. J Environ Sci 20:1281–1287
106. Yang N, Zhang H, Shao LM, Lü F, He PJ (2013) Greenhouse gas emissions during MSW landfilling in China: influence of waste characteristics and LFG treatment measures. J Environ Manage 129:510–521
107. Yang Y (2014) Economic compensation for waste incineration plants. Finance Econ 24:157–159 (in Chinese with English abstract)
108. Yeh LT (2020) Analysis of the dynamic electricity revenue inefficiencies of Taiwan's municipal solid waste incineration plants using data envelopment analysis. Waste Manage 107:28–35
109. Yi J, Yao S, Wu X, Yao X (2014) Research on the monitoring system of the surrrounding environment and the problem of economic compensation for surrounding residents mear a waste incineration plant. Environ Eng 32:742–744 (in Chinese with English abstract)
110. Yigitcanlar T, Sabatini-Marques J, Da-Costa EM, Kamruzzaman M, Ioppolo G (2019) Stimulating technological innovation through incentives: perceptions of Australian and Brazilian firms. Technol Forecast Soc Chang 146:403–412

111. You M, Li S, Li D, Cao Q, Xu F (2020) Evolutionary game analysis of coal-mine enterprise internal safety inspection system in China based on system dynamics. Resour Policy 67
112. Yu H, Wang W, Yang BH, Li CF (2019) Evolutionary game analysis of the stress effect of cross-regional transfer of resource-exhausted enterprises. Complexity 2019(1):7652430
113. Yuan H, Shen LY, Hao JJL, Lu WS (2011) A model for cost–benefit analysis of construction and demolition waste management throughout the waste chain. Resour Conserv Recycl 55:604–612
114. Zeng C, Niu D, Zhao Y (2015) A comprehensive overview of rural solid waste management in China. Front Environ Sci Eng 9(6):949–961
115. Zhang DQ, Tan SK, Gersberg RM (2010) Municipal solid waste management in China: status, problems and challenges. J Environ Manage 91(8):1623–1633
116. Zhang L, Klenosky DB (2016) Residents' perceptions and attitudes toward waste treatment facility sites and their possible conversion: a literature review. Urban Forestry Urban Greening 20:32–42
117. Zhang L, Long R, Huang Z, Li W, Wei J (2020) Evolutionary game analysis on the implementation of subsidy policy for sustainable transportation development. J Clean Prod 267:122159
118. Zhao R, Huang T, McGuire M (2012) From a literature review to an alternative treatment system for landfill gas and leachate. Challenges 3:278–289
119. Zhao R, Neighbour G, Deutz P, McGuire M (2012) Materials selection for cleaner production: an environmental evaluation approach. Mater Des 37:429–434
120. Zhao R, Neighbour G, McGuire M, Deutz P (2013) A software based simulation for cleaner production: a game between manufacturers and government. J Loss Prev Process Ind 26(1):59–67
121. Zhao R, Zhou X, Han J, Liu C (2016) For the sustainable performance of the carbon reduction labeling policies under an evolutionary game simulation. Technol Forecast Soc Chang 112(1):262–274
122. Zhao R, Xi B, Liu Y, Su J, Liu S (2017) Economic potential of leachate evaporation by using landfill gas: a system dynamics approach. Resour Conserv Recycl 124:74–84
123. Zhao R, Liu Y, Zhang N, Huang T (2017) An optimization model for green supply chain management by using a big data analytic approach. J Clean Prod 142:1085–1097
124. Zhao R, Han J, Zhong S, Huang Y (2018) Interaction between enterprises and consumers in a market of carbon-labeled products: a game theoretical analysis. Environ Sci Pollut Res 25(2):1394–1404
125. Zulkepli NE, Muis ZA, Mahmood NAN, Hashim H, Ho WS (2017) Cost benefit analysis of composting and anaerobic digestion in a community: a review. Chem Eng Trans 56:1777–1782
126. Marticorena B, Attal A, Camacho P, Manem J, Hesnault D, Salmon P (1993) Prediction rules for biogas valorization in municipal solid waste landfills. Water Sci Technol 27(2): 235-241.

Chapter 6
Summary and Future Study

6.1 Summary

Many of the materials currently used in production could be considered hazardous at any stage of their life cycle (i.e. from the time of extraction to final disposal); this may pose a potential risk to human health or the environment due to the quantity, concentration, or physical or chemical properties of the material in question [9, 13]. The inherent risk of improperly managing or disposing of hazardous materials can have a substantial impact on supply chain management, resulting in business losses, environmental pollution, and even casualties [5, 7, 12]. In such cases, consideration should be given to reducing the inherent risks associated with supply chain involving the hazardous materials in order to improve the social performance of sustainable production and consumption.

Climate change has been identified as another critical factor that strongly influences the performance of sustainable production and consumption, and can lead to significant uncertainty in terms of public acceptance of different products [6, 10]. For instance, consumer's purchasing preferences may be influenced by the emissions information provided on a product label, especially as they become more aware of climate change [1, 13, 15]. Carbon management has thus received increasing attention from a number of upstream and downstream enterprises in the supply chain, which are working to reduce their life cycle based carbon emissions and move towards a low-carbon business model [2, 3, 10, 11]. However, when making business decisions, the enterprises involved may still be driven by profitability, rather than prioritising risk control and carbon emission reduction because of the additional expenditure and high financial cost of such environmental protection efforts [4, 8, 14]. Thus, understanding how to reduce environmental risks and carbon emissions without compromising economic sustainability is a critical challenge for proponents of sustainable consumption and production.

R. Zhao, *Multi-criteria Decision-Making Approaches to Sustainable Consumption and Production*, Urban Sustainability, https://doi.org/10.1007/978-981-96-2775-2_6

This book focuses on the application of multi-criteria decision making for sustainable consumption and production to minimize the associated impacts, i.e. environmental risks and carbon emissions, thereby providing novel solutions for societal change in the context of sustainable development. In particular, the book aims to promote sustainable consumption and production while maintaining industrial competitiveness, and to demonstrate the sustainability of environmentally friendly products, taking the Triple Bottom Line into account. The summary of the book and the recommendations for future study are given below.

Chapter 1 provides a brief background on the research topic and then discusses the problem situation that is essential to propose multi-criteria decision making for sustainable consumption and production. Chapter 2 considers the inherent risk of hazardous materials, their associated carbon emissions, and total economic costs as the main objectives that could be minimized to improve the eco-efficiency of supply chain management. Chapter 3 provides a novel perspective that models the interactions of possible stakeholders involved in sustainable consumption and production, in response to increasing performance along the three dimensions of the Triple Bottom Line. Chapter 4 mainly focuses on multi-criteria decision making to aid materials selection by taking sustainable design into account. Chapter 5 shows decision support for waste management by seeking the optimal or near-optimal strategies for decreasing environmental risks and greenhouse gases emissions to achieve sustainable waste management.

6.2 Future Study

The book offers an approach to maintaining competitiveness and demonstrating sustainability for products, i.e. economic, social and environmental dimensions. However, there are still some limitations in the presented study, which lay a foundation for future work to provide appropriate strategies for promotion of sustainable production and consumption.

With regard to the study on improving the eco-efficiency of supply chain management, the supply chain network in the book is assumed to be a single-period product supply chain. The risk defined in the study only focuses on the immediate risk of sudden accidents, which does not take long-term consequences into account. The impact that reducing the environmental risks and carbon emissions can have on the long-term economic performance of the supply chain network has yet to be investigated. In addition, the big data approach has been applied to data acquisition and data quality control in this study. However, there are very few studies regarding to the integration of big data science and supply chain management. Further study will center on improvement of the model, including considering the influence of stored and backordered inventory on the optimization of the supply chain network. This further study will be conducted using a system dynamics approach, and it will also consider the cumulative effects of inherent risk, and the long-term effects of environmental risks and carbon emissions on the costs incurred by the various stakeholders in adopting green supply chain management practices. With regard to the application of big data approach, it is expected that further research will better incorporate big data analysis models, such as panel data models, into the supply chain management.

There is room for future improvement in the analysis of possible interactions between the actors involved in sustainable consumption and production. First, the manufacturers are considered as a single entity, regardless of the scale, quantity and quality of production, economic benefits, strategies selected, etc. Second, in order to simplify the game model, the interactions between the upstream and downstream businesses, as well as between the enterprises and consumers, have been omitted. In addition, consumer preference for environmentally friendly products is assumed to be the main factor driving demand for better, safer and more sustainable products for consumption. However, the influence of factors such as living standards, public purchasing behaviour, etc. has not yet been considered in this study. Increased individual environmental awareness can lead to a shift in consumption patterns towards sustainability. Especially as low-carbon development has become a global common sense, consumer neuroscience may help to motivate pro-environmental behavior based on consumer market segmentation. Future studies will centre on improving the game model, including the differences between the manufacturers, the interactions between the upstream and downstream enterprises, manufacturers and consumers etc.

For material selection, the evaluation criteria are generic in the context of sustainable consumption and production. This situation can be represented as whether the selected indicators can be used to fully cover the three dimensions of sustainability, namely, the economic, environmental, and social dimensions. It can be further improved to include other typical parameters considered in the material selection process, e.g. physical and chemical properties of materials, material performance for the specific process industries, thus to make the evaluation results more convincing in different application scenarios. Besides, the weighting determination can be re-adjusted by taking the preferences of the engineering designer into account, while giving full priority to environmental and social considerations. It is expected that the material selection approaches proposed in this book will be applied to more case scenarios in order to test and refine their validity and sensitivity in the future.

The proposed planning scenario for sustainable waste management is limited in scope by the assumptions made about the claimed economic benefits of combining existing technologies and creating an integrated waste treatment park designed to overcome both the difficulties associated with landfill gas transport and the costs of leachate treatment. Furthermore, the capital costs associated with the treatment of gaseous emissions and the operational costs of ash disposal were not included in the analysis. Future work will consider improvements to the development of the disposal scenario, in particular a more detailed consideration of technical and economic feasibility.

References

1. Abdallah T, Farhat A, Diabat A, Kennedy S (2012) Green supply chains with carbon trading and environmental sourcing: formulation and life cycle assessment. Appl Math Model 36:4271–4285
2. Acquaye AA, Wiedmann T, Feng K, Crawford RH, Barrett J, Kuylenstierna J, McQueen-Mason S (2011) Identificationof 'carbon hot-spots' and quantification of GHG intensities in the biodiesel supply chain using hybrid LCA and structural path analysis. Environ Sci Technol 45:2471–2478
3. Benjaafar S, Li Y, Daskin M (2013) Carbon footprint and the management of supply chains: insights from simple models. IEEE Trans Autom Sci Eng 10:99–116
4. Govindan K, Kaliyan M, Kannan D, Haq AN (2014) Barriers analysis for green supply chain management implementation in Indian industries using analytic hierarchy process. Int J Prod Econ 147:555–568
5. He G, Zhang L, Lu Y, Mol AP (2011) Managing major chemical accidents in China: towards effective risk information. J Hazard Mater 187:171–181
6. Lee KH (2011) Integrating carbon footprint into supply chain management: the case of Hyundai Motor Company (HMC) in the automobile industry. J Clean Prod 19:1216–1223
7. Mangla SK, Kumar P, Barua MK (2014) A flexible decision framework for building risk mitigation strategies in green supply chain using SAP–LAP and IRP approaches. Glob J Flex Syst Manage 15:203–218
8. Mollenkopf D, Stolze H, Tate WL, Ueltschy M (2010) Green, lean, and global supply chains. Int J Phys Distrib Logist Manag 40:14–41
9. Pradhananga R, Taniguchi E, Yamada T, Qureshi AG (2014) Bi-objective decision support system for routing and scheduling of hazardous materials. Socioecon Plann Sci 48:135–148
10. Singh A, Mishra N, Ali SI, Shukla N, Shankar R (2015) Cloud computing technology: reducing carbon footprint in beef supply chain. Int J Prod Econ 164:462–471
11. Sundarakani B, De Souza R, Goh M, Wagner SM, Manikandan S (2010) Modeling carbon footprints across the supply chain. Int J Prod Econ 128:43–50
12. Wang X, Chan HK, Yee RW, Diaz-Rainey I (2012) A two-stage fuzzy-AHP model for risk assessment of implementing green initiatives in the fashion supply chain. Int J Prod Econ 135:595–606
13. Zhao R, Neighbour G, Han J, McGuire M, Deutz P (2012) Using game theory to describe strategy selection for environmental risk and carbon emissions reduction in the green supply chain. J Loss Prev Process Ind 25:927–936
14. Zhao R, Neighbour G, McGuire M, Deutz P (2013) A software based simulation for cleaner production: a game between manufacturers and government. J Loss Prev Process Ind 26(1):59–67
15. Zhao R, Zhong S (2015) Carbon labelling influences on consumers' behaviour: a system dynamics approach. Ecol Ind 51:98–106

The manufacturer's authorised representative in the EU is Springer Nature Customer Service Centre GmbH, Europaplatz 3, 69115 Heidelberg, Germany. If you have any concerns regarding our products, please contact ProductSafety@springernature.com

Printed and bound by CPI Group (UK) Ltd, Croydon, CR0 4YY
30/09/2026
02225300-0005